56 Topics in Current Chemistry

Fortschritte der chemischen Forschung

W0106088

Theoretical Inorganic Chemistry

Springer-Verlag
Berlin Heidelberg GmbH 1975

This series presents critical reviews of the present position and future trends in modern chemical research. It is addressed to all research and industrial chemists who wish to keep abreast of advances in their subject.

As a rule, contributions are specially commissioned. The editors and publishers will, however, always be pleased to receive suggestions and supplementary information. Papers are accepted for "Topics in Current Chemistry" in either German or English.

ISBN 978-3-662-15948-4

Originally published by Springer-Verlag Berlin Heidelberg New York in 1975
Softcover reprint of the hardcover 1st edition 1975

Library of Congress Cataloging in Publication Data. Main entry under title: Theoretical inorganic chemistry. (Topics in current chemistry; 56). Bibliography: p. Includes index. CONTENTS: Jørgensen, C. K. Continuum effects indicated by hard and soft antibases (Lewis acids) and bases. — Brunner, H. Stereochemistry of the reactions of optically active organometallic transition metal compounds. [etc.]. 1. Chemistry, Physical and theoretical – Addresses, essays, lectures. I. Series.
QD1.F58   vol. 56 [QD455]   540'.8s [546]   75-5565
ISBN 978-3-662-15948-4   ISBN 978-3-540-37558-6 (eBook)
DOI 10.1007/978-3-540-37558-6

Contents

Continuum Effects Indicated by Hard and Soft Anti-bases (Lewis Acids) and Bases

Professor Dr. Christian Klixbüll Jørgensen

Département de Chimie minérale et analytique, Université de Genève, CH-1211 Geneva 4, Switzerland

Contents

1

I. Historical Introduction

A. Chemical Affinity

The main object of classical physical chemistry (to be distinguished from chemical physics studying atomic and molecular properties) is to clarify the relation between *free energy G* and *enthalpy H*. By obvious analogy to combustibles, it was felt that the characteristic feature of chemical bonding is the evolution of heat. Around 1880 when Julius Thomsen and other chemists performed very precise calorimetric measurements, it was argued that the chemical avidity was essentially the heat evolution with possible small correction terms. However, one could not continue to ignore that many spontaneous reactions are strongly endothermic such as the dissolution of ammonium nitrate in water, and the majority of chemical equilibria are not calorimetrically neutral to the extent that the reaction progressing in one direction is exothermic, but when the original concentrations are on the other side of equilibrium, the opposite reaction should turn out also to be exothermic. It was realized that physicochemical equilibria are determined by a competition between heat evolution and a tendency toward disorder, favouring the solution compared with the crystals and pure water, or for that matter the dilute vapour formed by endothermic evaporation of the liquid. The quantitative measure of disorder is *entropy S* representing the product of the gas constant R and the natural logarithm of the number of accessible microstates (assuming axiomatic equal probability of each microstate, this number is proportional to the probability of the macroscopic state) in statistical mechanics. For a chemical reaction, the important relation is

$$\Delta G = \Delta H - T\Delta S \qquad (1)$$

choosing the sign in such a way that spontaneous reactions have negative ΔG and exothermic reactions negative ΔH.

Guldberg and Wåge suggested the mass-action law in 1863 based on kinetic arguments. A much more clear-cut derivation is possible from the hypothesis that the free energy contains a contribution from the natural logarithm of the concentration C of the substance:

$$G = G_0 + RT \ln C \qquad (2)$$

The zero-point G_0 depends on the solvent and conceivably on other variables. Thus, the distribution coefficient K_d between two phases correspond to the species being $\exp[(G_{01} - G_{02})/RT]$ times more soluble in the phase 2 than in the phase 1. The solubility product K_{sp} is an expression of G_0 in the solid compared with G_0 in the solution, if we accept to speak about free energies of ions and not only of neutral species. Since Arrhenius convinced his reluctant contemporaries about the predominant (or complete) dissociation of salts in aqueous solution, it became acceptable to write complex formation constants involving concentrations of ions and not only of neutral molecules.

It is worthwhile to discuss why the mass-action law on *concentration basis* (moles/litre) is plausible. It is beyond doubt that it is not always valid. The concentration 5.5 M of saturated aqueous sodium chloride indicates the solubility product 30 moles^2/litre^2. If an equal amount of such a solution is added to 12 M hydrochloric acid, the concentration of Na$^+$ is 2.75 M and of Cl$^-$ $(12 + 5.5)/2 = 8.75$ M. Their product 24.06 M^2 is distinctly below the solubility product, but nevertheless, more than 80 percent of the NaCl present crystallizes out. It would be to short-circuit this paradox to speak about the mass-action law on activity basis. The introduction of activity a as the product $a = fC$ of the activity coefficient f and the concentration is a tautological trick to keep the mass-action law valid, and it is more fruitful to try to explain why f varies more dramatically in some cases than in others.

It is an empirical fact that mixtures of most gases at pressures below a few atmospheres are *almost ideal* following a set of rules common to all gases. Exceptions must be made for mixtures precipitating condensed matter (liquids, amorphous-vitreous and crystalline solids) not remaining gaseous, *e.g.* NH$_4$Cl from NH$_3$ and HCl at temperatures below some 350 °C, or more extensive reactions such as SO$_2$ and Cl$_2$ forming SO$_2$Cl$_2$. In normal cases, the partial pressures of each constituent are remarkably additive. It is not always recognized that the law of Boyle and Mariotte

$$PV = nRT \qquad (3)$$

contains the hypothesis of Avogadro in the determination of the number n of moles. This hypothesis was not generally accepted before the Congress in Karlsruhe 1860 and has some unexpected corollaries such as the oligo-atomicity of gaseous elements such as H$_2$, N$_2$, O$_2$ (and ozone O$_3$), P$_4$ and S$_8$ (in equilibrium with other species such as S$_2$ and S$_6$). It is well-known that chemical equivalent weights cannot be convincingly multiplied to atomic weights without such an assumption, and sometimes troubles occured. The chemical properties of beryllium are much more similar to those of aluminium than to magnesium and calcium, and there were serious suspicions that Be is trivalent. Metallic beryllium happened to be one of the few cases of strong deviations from Dulong and Petit's law (that the specific heat of a metal is 6 gcal/°C per mole, by the way also valid for the average atomic weight 6 of water) suggesting the atomic weight 13 rather than 9. To make matters worse, the vapour of beryllium chloride contains appreciable amounts of the dimer ClBeCl$_2$BeCl again supporting trivalency. Comparable cases, such as the loosely bound colourless dimer formed at lower temperature from the brown monomer NO$_2$ represent deviations from Boyle and Mariotte's law, in sofar Eq. (3) is not satisfied when varying the absolute temperature T. It is not always easy to predict whether a given gaseous compound tends toward oligomerization. Whereas HF contains a large proportion of (HF)$_2$ and (HF)$_6$ pure HCl and NH$_3$ are almost ideal gases, and acetic acid vapour contains some hydrogen-bonded dimers. The situation is strikingly different in aqueous solution. Though HF reacts with F$^-$ to form FHF$^-$, the equilibrium between acetic acid and acetate is almost a Brønsted monomeric acid-base pair (large amounts of dimerized acetic acid in solution would increase pK at higher concentration) and aqueous ammonia is not exceptionally deviating from ideality. At 25 °C, a solution 16 M NH$_3$ (29 M H$_2$O) has the ammonia pressure 1 atm, and pure liquid ammonia

(36 M) 10 atm to be compared with 0.02 atm for 1 M aqueous ammonia. It is a well-known fact that dilute HCl solutions smell to a much smaller extent than dilute NH_3. Though 1 atm pressure occurs at 25 °C in a solution (13 M HCl, 40 M H_2O) not extremely different in molarity from ammonia, pure liquid hydrogen chloride (22 M) is 47 atm and 6 M HCl is already close to the constant-boiling hydrochloric acid used for titrimetric standardization, the HCl vapour pressure 0.0004 atm (to be compared with 10^{-5} atm for 3 M HCl) clearly showing a strong deviation from ideality. 1 M HCl has the HCl pressure 100000 times lower than that of H_2O. It can be argued that the decrease of the activity coefficient f by a factor $1.2 \cdot 10^{-11}$ (22/47) relative to liquid HCl is an expression of a very low pK (~ -10) of HCl as a Brønsted acid, reacting almost quantitatively to H_{aq}^+ and Cl^- in dilute aqueous solution (where pK of H_{aq}^+ would be $-\log 55 = -1.74$ taking the molarity 55 of water into account).

Anyhow, at 25 °C, an ideal gas at 1 atm is 0.041 M. Condensed matter with small molecules (or metals such as silver and gold) can be up to 100 M. Hence, at their boiling points, most substances show an activity coefficient in the gaseous state (comparing with the molarity of the condensed matter and not the conventional activity $a = 1$ of pure substances) of the order of magnitude 1000. In view of the almost ideal nature of the gaseous state, it would perhaps be more appropriate to say that the condensed matter has $f \sim 10^{-3}$ relative to the vapour at 1 atm.

It is possible to discuss solvents almost immiscible with water along this line. 1 litre benzene can dissolve 700 mg water. Such a saturated solution has the same partial pressure of H_2O as pure water. Since the two molarities are 0.04 and 55 M, the activity coefficient f (on a molarity basis) is 1400 for water in C_6H_6. This large value can be made even larger for solvents in which water is even less soluble. Suppose 1 litre carbon tetrachloride can dissolve 180 mg H_2O forming a 10^{-2} M solution. Then, $f = 5500$. It is not without interest to compare with saturated water vapour at 25 °C being 0.0012 M. Said in other words, carbon tetrachloride is a 8 times better solvent than empty space for water. This statement looks pretty ludicrous to the professional physico-chemist because empty space is not normally considered as a solvent but is rather considered as formed *ad libitum* by moving the walls confining the system studied. However, in view of the approximate ideality of dilute gases, the solubility of water in the system 0.04 M argon, or the more frequent system 0.032 M N_2 and 0.008 M O_2, is also close to 0.0012 M, and it is not unreasonable to extrapolate this statement to empty space through intermediate stages such as 10^{-7} M helium. There is a clear analogy between the distribution between two solvents and between one solvent and empty space. In the latter case, $K_d = 5 \cdot 10^4$ for water and $K_d = 7 \cdot 10^8$ for mercury at 25 °C. Almost the same value $7 \cdot 10^8$ is obtained for the solubility of mercury atoms in water.

However, the idea of a chemical potential containing a term proportional to the logarithm of the concentration like Eq. (2) soon gained confidence for ions. Nernst's law for electrode potentials varying (at 25 °C) 0.059 V divided by the number of electrons removed in the oxidation process for each power of ten in the concentration of free (*i.e.* exclusively hydrated) metal ions in solution soon got used for estimating ion concentrations below 10^{-20} M having no analytical significance but rather representing exponential functions of free energy differences like Eq. (2).

By the same token, it originally demanded an enormous experimental effort to purify water by distillation and keep it free from contaminants such as CO_2 in order to prove that the pure substance indeed contains 10^{-7} M H_{aq}^+ and 10^{-7} M OH^- in equilibrium. After S. P. L. Sørensen introduced the concept pH, it became plausible to determine pH by electrode potentials, first from the classical hydrogen electrode with finely divided platinum catalyst (the conditions for the standard oxidation potential E^0 to show its zero point is a monument for the difficulties when mixing thermodynamical prescriptions for gaseous and condensed matter: 0.041 M H_2 in gas at 25 °C, 1 M H_{aq}^+ and roughly 54 M H_2O) and later from oxidized antimony, from quinhydrone, and from better and better glass electrodes, all being easier to handle but more difficult to understand as far goes the detailed mechanism.

One of the most helpful things to realize when applying thermodynamic arguments to reactions in solution is that usually *some but not all* reactions achieve equilibrium under the conditions prevailing in the laboratory at the time-scale of the measurements. Thus, when determining pK close to 9 of Fe $(H_2O)_6^{+2}$ as a Brønsted acid in a perchlorate solution, it is considered irrelevant that ClO_4^- could oxidize eight Fe(II) to Fe(III) (of which the hexaaqua ion has pK $\sim$ 3) whereas it is not without experimental importance to exclude oxygen carefully. On the other hand, strong solutions of hydrochloric acid are normally handled without considering the thermodynamically feasible formation of two Cl_2 for each O_2. Electron-transfer (redox) reactions are frequently slow and may need rather odd catalysts, whereas Brønsted acidity (with exception of some C—H bonded cases, such as the deprotonation of nitromethane) normally is rapid. However, this is only true for monomeric acid-base pairs; oligomeric hydroxo complexes of trivalent aluminium, chromium, iron, rhodium and bismuth and of quadrivalent tin, cerium and hafnium form slowly, and react slowly with excess acid, as is also true for pyrophosphate $O_3POPO_3^{-4}$ but not for the analogous dichromate. In principle, one should not be able in aqueous solution (say above 40 M H_2O) to get outside the parallelogram in a (E^0, pH) plot limited by the vertical lines pH = -1 and $+15$ and the two slanted, straight lines $E^0 = (1.23 - 0.059 \text{ pH})$ V and $E^0 = (-0.059 \text{ pH})$ V. Empirically, the opportunities for oxygen overvoltage frequently allow solutions to be situated above the former slope and of hydrogen overvoltage to be below the latter line.

B. Determination of Complex Formation Constants

Around 1900, Abegg and Bodländer performed measurements of copper, silver and mercury electrode potentials which could be interpreted as the formation of a definite complex with an excess of a strongly bound ligand. The choice of efficient metallic electrodes selected several cases of the coordination number $N = 2$ such as Ag $(NH_3)_2^+$ in an excess of ammonia or Ag $(CN)_2^-$ in an excess of cyanide. It was noted that a few compounds dissolve as neutral molecules (like sugars or urea) in water without providing much electric conductivity, such as $HgCl_2$ and Hg $(CN)_2$ now known to be triatomic and penta-atomic linear molecules. However, in an excess of Cl^- or CN^-, the ionic species $HgCl_4^{-2}$ and Hg $(CN)_4^{-2}$ are formed. These measurements were performed on solutions establishing equilibrium very quickly.

5

On the other hand, the complexes of cobalt(III), rhodium(III), iridium(III) and platinum(IV) (all with $N = 6$) and of platinum(II) ($N = 4$) studied in the last century by Blomstrand and S. M. Jørgensen were not amenable to equilibrium studies though Alfred Werner[1] generalized the idea of an (almost invariant) coordination number N for an element in a definite oxidation state. It is remarkable that the corresponding geometrical models of octahedra, squares, tetrahedra, . . . were made before the first crystal structures were determined by diffraction of X-rays. Niels Bjerrum found the connecting link between preparative and formation constant-determining complex chemistry in the chromium(III) thiocyanate complexes with $N = 6$ of the general formula $Cr(NCS)_n(H_2O)_{6-n}^{+3-n}$ where all seven complexes react sufficiently slowly to be separated by chemical techniques, but which reach equilibrium at room temperature within a few days in such a way that the *formation constants*

$$K_1 = \frac{[ML]}{[M][L]} \qquad K_2 = \frac{[ML_2]}{[ML][L]} \qquad \ldots K_n = \frac{[ML_n]}{[ML_{n-1}][L]} \qquad (4)$$

$$\beta_2 = K_1 K_2 \qquad \beta_3 = K_1 K_2 K_3 \qquad \ldots \qquad \beta_n = K_1 K_2 \ldots K_n$$

could be determined. Originally, there was a tendency to give the reciprocal values $(1/K_n)$ as instability constants. For instance, it was known that gaseous PCl_5 dissociates to a certain extent to PCl_3 and Cl_2 but that the concomitant deviation from Avogadro's hypothesis can be suppressed by an excess of one of the two dissociation products. The brackets in the dissociation constant $[PCl_3][Cl_2]/[PCl_5]$ indicate molar concentrations (which are proportional to the partial pressures in ideal gas mixtures at a given temperature). Russian authors[2] still use instability constants whereas the large catalogs "Stability Constants"[3] use the definitions in Eq. (4).

One important aspect of Niels Bjerrum's studies is the recognition of the aqua ion $Cr(H_2O)_6^{+3}$ as a complex on equal footing with other complexes of neutral ligands such as $Cr(NH_3)_6^{+3}$. Nevertheless, it is the general agreement between determiners of formation constants to neglect the water of constitution and to put the activity of water = 1 and not 55 M. The main reason for this decision was that the number of water molecules previously only was known in very few cases. Even today, studies of visible absorption spectra of aqua ions containing partly filled d shells[4] and of proton and oxygen 17 nuclear magnetic resonance, and in a few instances Raman spectra, have allowed tetrahedral $Be(H_2O)_4^{+2}$, quadratic $Pd(H_2O)_4^{+2}$ and octahedral hexa-aqua ions of Mg(II), Al(III), Ti(III), V(II), V(III) Cr(III), Mn(II), Fe(II), Fe(III), Co(II), Co(III), Ni(II), Zn(II), Ga(III), Ru(III) and Rh(III) to be identified. In other cases, such as aquated Li^+, Ca^{+2}, Ag^+, Tl^+, Pb^{+2} and Th^{+4} it is by no means certain that a unique species occurs, and a statistical mixture of weakly bound non-equivalent water ligands is quite conceivable. It is fairly certain that Sc(III), Y(III) and the trivalent lanthanides occur as ennea-aqua ions, but if cerium(III) solutions contain a small quantity of $Ce(H_2O)_8^{+3}$ and a large quantity of $Ce(H_2O)_9^{+3}$, it may be possible to see spectral differences, but for the thermodynamicist the mixture just represents "the standard state of aqua ions in dilute solution". We shall see below that even more intricate problems occur when comparing the instantaneous picture (derived from spectra) of copper(II) aqua ions with crystal structures and other evidence related to time-average pictures. This does not prevent that the kinetic properties of chromium(III) complexes clearly show that the concentration of $Cr(H_2O)_5^{+3}$, a typical kinetic intermediate, in equilibrium is smaller than 10^{-15} times the concentration of the hexa-aqua ion. This comparison shows that a hypothetical species like unsolvated Cr^{+3} is completely unrealistic, exactly like unsolvated H^+. Steric hindrance between neighbour ligands, and stereochemical preferences determined by the electronic structure of the central atom, also prevents higher N from occuring, and the upper limit to the concentration of $Cr(H_2O)_7^{+3}$ with roughly equivalent ligands is even lower, because H. Taube otherwise would have found oxygen 18 exchange with the solvent more rapid than one day half-life.

In 1932, Jannik Bjerrum[5] started work on ammonia complexes of copper(I) and Cu(II). Some of the results were obtained with more direct techniques, such as measurement of the ammonia vapour pressure over the solution, or solubility of weakly soluble salts in various supernatant solutions. However, most of the work involved determination of the free concentration of ammonia [NH_3] via measurements with a glass electrode of

$$pH = 9.2 + \log [NH_3] - \log [NH_4^+] \tag{5}$$

where the ammonium ion concentration [NH_4^+] is very close to the total concentration of an ammonium salt (say 1 or 2 molar) which serves at the same time as a *constant salt medium*. This has to a certain extent been the satisfactory response to the worries about activity coefficients. Experience has shown that the effect of adding 0.1 or even 0.01 M of salt to pure water are larger and far more unpredictable than the effect of modifying, say 1 M $NaClO_4$, to 0.9 M sodium perchlorate and 0.1 M of another salt. Anyhow, one has to be careful when applying Eq. (5) which only indicates the consumption of ammonia as the difference between the total concentration of ammonia entering the system and [NH_3]. For instance, aqua ions may very well deprotonate to hydroxo complexes by transforming NH_3 to NH_4^+. The distinction between genuine ammonia complexes and such parasitic ammonia removal can be made by varying the total concentration of NH_4^+ (say by a factor of 2 or 5) if the mass-action law is valid. Later, J. Bjerrum[6] succeeded in determining $\log K_6 = 4.4$ of the cobalt(III) ammonia system by careful study of

$$Co(NH_3)_5OH^{+2} + NH_4^+ = Co(NH_3)_6^{+3} + H_2O \tag{6}$$

in the presence of active charcoal as a catalyst and knowing pK = 6.5 of the aqua ion $Co(NH_3)_5(H_2O)^{+3}$ as a Brønsted acid. The product of the six constants are known to be $\log \beta_6 = 35.2$ for the cobalt(III) and 5.2 for the cobalt(II) ammonia complexes. The difference between the two values of $\log \beta_6$ can be determined from the change of the standard oxidation potential E^0 from + 1.85 V for $Co(H_2O)_6^{+2}$ to + 0.1 V for $Co(NH_3)_6^{+2}$ which is much easier to oxidize. Actually, 1.75/0.059 = 30.

In the case of copper(I) ammonia complexes, $Cu(NH_3)_2^+$ is formed with $\log \beta_2$ = 10.9 and higher complexes cannot be detected, in close analogy to the silver(I) $Ag(NH_3)_2^+$ having $\log \beta_2 = 7.0$. On the other hand, copper(II) turned out to be considerably more complicated. It is true that between 0.08 and 0.3 M NH_3, more than

90 percent occurs as $Cu(NH_3)_4^{+2}$ with $\log K_4 = 2.1$ and $\log \beta_4 = 12.7$ to be compared with $Cu(NH_3)(H_2O)_x^{+2}$ having $\log K_1 = 4.1$. These facts give a quantitative description of the reaction

$$Cu + Cu(NH_3)_4^{+2} = 2\,Cu(NH_3)_2^{+} \tag{7}$$

progressing extensively to the right in the absence of oxygen and other oxidants. On the other hand, copper(I) aqua ions disproportionate to metallic copper and Cu(II) except at very low concentrations, because the activity of the element is unity disregarding the amount of solid.

However, a fifth ammonia[5] is bound in $Cu(NH_3)_5^{+2}$ with $K_5 = 0.3$, *i.e.* an equal amount of tetrammine and pentammine occurs in 3.3 M NH_3. The pentammine is formed with a characteristic change of the visible absorption spectrum, and its constitution is further discussed below. Jannik Bjerrum[6] points out that complexes containing only one central atom M (in contrast to many oligomeric hydroxo complexes) assuming the mass-action law has an average number of ligands bound per M

$$\bar{n} = ([ML] + 2\,[ML_2] + \ldots + n\,[ML_n])/([M] + [ML] + \ldots + [ML_n]) \tag{8}$$

not depending on the total M concentration but only on [L]. J. Bjerrum calls the plot of $\bar{n}$ as a function of $\log[L]$ the *formation curve*. He introduced the distinction between the *maximum coordination number* $N_{max} = 5$ in $Cu(NH_3)_5^{+2}$ and 4 in the mercury(II) complexes discussed above, and the *characteristic coordination number* $N_{char} = 4$ in Cu(II) and 2 in Hg(II). J. Bjerrum also discussed deviations from the statistical behaviour in the intermediate complexes, which are normally mixed complexes of the unidentate (neutral or charged) ligand L^{-y} and water having the constitution $ML_n(H_2O)_{N-n}^{+z-ny}$. In the purely statistical distribution of ligands on N available sites, an average constant K_{av} is defined by $\log \beta_N = N \log K_{av}$, and $K_1 = NK_{av}$ whereas $K_N = K_{av}/N$. In the simplest case of statistical distribution on two sites, $K_1 = 4K_2$. The formation curve has the simple form

$$\bar{n} = NK_{av}[L]/(1 + K_{av}[L]) \tag{9}$$

in the statistical case of N equivalent sites. Usually, the ratio (K_{n+1}/K_n) is somewhat smaller than the statistical value $n(N-n)/(Nn - n^2 + N + 1)$. In such cases it is a good approximation to K_{n+1} to take the reciprocal concentration $1/[L]$ at $\bar{n} = n + 1/2$.

The approximate statistical distribution of unidentate ligands on N sites is a special case of the hypothesis of *step-wise complex formation* that all the intermediate complexes ML_n occur in mixtures with $0 < \bar{n} < N$. This hypothesis has turned out to be a remarkably successful alternative to the distinction by Job between *perfect* (robust, preparatively separate) and *imperfect* (labile, equilibrating in solution) complexes. The latter suggestion is similar to the influence of Aristotle's principle of the excluded middle on a housewife when she declares that a given compound either is toxic or not. Though most of the reactions of iridium(III) are slower than the reactions of carbon compounds in organic chemistry, there is little doubt that most mixtures of two ligands (including the case of one ligand and water) or three or more ligands do indeed contain non-negligible concentrations of all the mixed complexes of a given central atom. It is true, however, that some practical situations can raise some doubts about the universal validity of the hypothesis of step-wise formation. For instance, the cyanide complexes of iron(II), iron(III) and nickel(II) show almost exclusively $n = N$ (or $N - 1$ in a few cases) in homogeneous solution because of a combination of two important effects preventing other inter-mediate complexes from being detected: the lower n values form insoluble precipi-tates, and the stability can change suddenly when the total spin quantum number S of the groundstate decreases by one or two units when n increases. On the other hand, the spectra of all the intermediate chromium(III) cyanide complexes $Cr(CN)_n(H_2O)_{6-n}^{+3-n}$ have been studied[7] including the two geometrical isomers for each of the values $n = 2, 3$ and 4. Hence, this system is as well characterized as the rhodium(III) chloro complexes $RhCl_n(H_2O)_{6-n}^{+3-n}$ where equilibria can be obtained[8] by heating for several days at 120 °C. At room temperature, the individual species can be separated on ion-exchange resins much in the same way as the ten species (including geometrical isomers) $MX_nY_{6-n}^{-2}$ can be separated by electrophoresis[9, 10] of osmium(IV) chloro-bromo, chloro-iodo and bromo-iodo mixed complexes, as well as iridium(IV) chloro-bromo complexes.

Several books[11-17] treat the detailed techniques of how to determine forma-tion constants. In practice, classical physico-chemical measurements of one parameter, such as pH or $[M^{+z}]$ from an electrode potential, continue to furnish a large number of the constants published[3] though in principle "fingerprint techniques" such as visible spectra at a large number of wave-lengths, infra-red and Raman vibrational spectra, nuclear magnetic resonance without time-averaged ligand exchange, and rapid evaporation in mass spectrometers should be able to detect several complexes simultaneously. It is by no means a trivial question to ask for the constitution of the complexes for which the formation constants are sought, and Lars-Gunnar Sillén attempted to write an objective computer programme "Letagrop" producing estimates of accuracy and negligible or slightly negative formation constants for complexes not actually existing in significant amounts. Nevertheless, some of the oligomeric hydroxo complexes investigated in Stockholm provided unexpected predominant species such as $Al_{13}(OH)_{32}^{+7}$ (the crystal structure favours the version $AlO_4Al_{12}(OH)_{24}^{+7}$ as an iso-poly-cation) and octahedral $Bi_6(OH)_{12}^{+6}$ being hydrated, hexamerized BiO^+.

C. (A) and (B) Trends in Complex Chemistry

The natural history of chemistry consists since many centuries in empirical rules about the relative affinity of various elements. Though metathetical reactions are frequently used in preparation, the main origin is the filtering of insoluble precipi-tates, as when an aqueous solution of silver sulphate and another of barium iodide leave almost pure water in the filtrate:

$$2 Ag^+ + SO_4^{-2} + Ba^{+2} + 2 I^- = 2 AgI + BaSO_4 \qquad (10)$$

9

However, comparable exchange of neighbour atoms (*ligating* atoms) can occur in the gaseous state

$$BCl_3 + 6\ CH_3OH = B(OCH_3)_3 + 3\ CH_3Cl + 3\ H_2O \tag{11}$$

where the "driving force" seems to be the affinity of boron to oxygen atoms. It has been recognized for a long time (and is one of the consequences of the Avogadro hypothesis) that many highly exothermic reactions do not modify the total number of chemical bonds, in particular if the oxygen molecule is assumed to have a double bond:

$$CH_4 + 2\ O_2 = CO_2 + 2\ H_2O$$
$$2\ H_2 + O_2 = 2\ H_2O \tag{12}$$
$$H_2 + F_2 = 2\ HF$$

Hence, some hetero-atomic bonds must be considerably stronger than the homo-atomic bonds in the elements. Linus Pauling considered this tendency to be so fundamental that the defined the *thermochemical electronegativity* x from the relation of bond strengths (ΔH_0 values) D for dissociation to gaseous atoms

$$D_{AB} = \frac{1}{2}\ (D_{AA} + D_{BB}) + (\chi_A - \chi_B)^2\ eV \tag{13}$$

where the energy unit 1 eV = 23.05 kcal/mole = 8068 cm^{-1}. A corollary of Eq. (13) is that four atoms tend to rearrange according to AB + CD $\longrightarrow$ AD + BC in such a way that the most different χ occur for the two atoms in one of the products, say AD. This explanation is quite satisfactory for the "combustions" in Eq. (12) where χ of the atoms increases in the order H < C < O < F. However, less exothermic reactions do not always proceed according to the criterium that one of the products is as electrovalent as possible. As a matter of fact, Pearson[18] compared a large number of reactions of the type AB + CD and found exceptions to the rule of maximum χ difference in about a third of the cases. As a slightly extreme example, it cannot be argued that the pyrolysis in vacuo of barium azide Ba(N$_3$)$_2$ to metallic barium and nitrogen increases the tendency towards charge separation.

Besides the question whether one should consider ΔG or ΔH, it was early recognized that some elements, such as magnesium, aluminium, calcium, lanthanum and thorium, prefer fluoride to the heavier halides by precipitating insoluble fluorides, and other elements such as beryllium, boron, silicon, titanium, zirconium, niobium and tantalum prefer fluoride by forming highly complex anions in solution such as BeF_4^{-2}, BF_4^-, SiF_6^{-2}, TiF_6^{-2} etc. The opposite tendency, preferring iodide to bromide, and bromide to chloride, when forming insoluble halides or complex anions, is shown by copper(I), rhodium(III), palladium(II), silver(I), platinum(II), platinum(IV), gold(I), gold(III), mercury(II), thallium(III), lead(II) and bismuth(III). Abegg and Bodländer pointed out that the latter elements are *noble*, their E^0 relative to aqua ions in acidic solution (with exception of Tl and Pb) are positive (American chemists used for many years the opposite sign of E^0) corresponding to the metals not developing hydrogen with non-complexing acids (HI or HCN may do if the iodo or cyano complexes are very strong) but needing strong oxidants such as nitric acid to dissolve. On the other hand, the former group of metals preferring fluoride have negative E^0 evolving H_2 with acids or even with water alone.

It is not easy to tell whether it is an accident that noble metals prefer to form highly covalent complexes. It might be argued that one reason for high E^0 is a high heat of sublimation of the metal, but mercury is a distinct exception at this point, and tungsten with high boiling-point is rather of the fluoride-affine type.

The common technique of group separation in qualitative inorganic analysis was to use the highly varying solubilities of sulphides produced by saturated (about 0.1 M) H_2S at various pH. Since $pK_1 = 7$ and $pK_2 = 14$ of H_2S, the free sulphide concentration $[S^{-2}]$ is approximately 10^{-22} M at pH = 0, 10^{-14} M at pH = 4 and 10^{-6}M at pH = 9. It is perfectly clear that the former value is derived from Eq. (2) and is not a statement about one sulphide ion in 17 ml solution. On the whole, the iodide-fans precipitate sulphides at pH = 0 and the fluoride-addicts do not. Certain fringe cases such as cadmium(II) and lead(II) may need pH = 1 for precipitation as sulphides, but they are not particularly insisting on iodide either. Certain elements, such as arsenic, molybdenum and rhenium have the peculiar property of precipitating the sulphides most readily in strong hydrochloric acid (pH = −1) but less readily at positive pH. This behaviour may be connected with the amphoteric behaviour of these sulphides. Though the solution of sulphides in an excess of HS^- is known to contain thio-anions such as AsS_3^{-3} and SbS_4^{-3} it is quite possible that mixed oxo-thio complexes are formed here, in analogy to the known $ReOS_3^-$ [19]. As far goes As(III), Mo(VI) and Re(VII) there is no evidence available for a preference for iodide among the halides. However, mixed oxo-fluoro complexes may form in fluoride-containing solutions.

Goldschmidt introduced a classification of elements for *geochemical* purposes in *aerophilic* (noble gases and, to a certain extent, nitrogen), *thalassophilic* (mainly chlorine and bromine), *lithophilic* (the alkaline and alkaline-earth metals, aluminium, rare earths, silicon, zirconium, thorium, niobium, tantalum, etc.) *chalkophilic* (copper, zinc, arsenic, selenium, silver, cadmium, indium, antimony, tellurium, gold, mercury, thallium, lead, bismuth etc.) and *siderophilic* elements (iron, cobalt, nickel and also the six platinum group elements which would have been classified as chalkophilic on usual chemical evidence). The five types of elements would concentrate (by a kind of extraction process on a geological time-scale) in the atmosphere, the ocean, in silicate rocks, in sulphur-containing minerals and in the iron core known to give the Earth its high average density. The fifth category was constructed mainly on basis of the composition of the meteorites close to rust-free steel, and cannot be considered as a thermodynamical equilibrium, since the coinage metals would be extracted in iron. A practical consequence of this distribution is that the main rocks are mixed oxides of silicon and the lithophilic metals. As already Goldschmidt pointed out, such minerals can show non-stoichiometric composition with *charge compensation* as when the ratio between sodium(I) and calcium(II) is the same as the ratio between silicon(IV) and aluminium(III), or when monazite $LnPO_4$ being the phosphate of the mixed lanthanides Ln has some La(III) replaced by Th(IV) and simultaneously the same amount of P(V) replaced by Si(IV). In other cases, mixed oxides can be non-stoichiometric by oxide deficit[20] such as $Ln_x Th_{1-x} O_{2-0.5x}$ or the Nernst lamp $Ln_x Zr_{1-x} O_{2-0.5x}$ conducting at high temperature by transport of oxide (rather than of electrons, as magnetite Fe_3O_4 does). In such cases, x may vary continuously from zero to above 0.5. It is characteristic for substitution of cations in mixed oxides that comparable ionic radii is a more important condition than identical oxidation states. Thus, one can easily introduce some ($x \sim 0.1$) magnesium(II) in $Mg_x Zr_{1-x} O_{2-x}$.

Since there is much less sulphur than oxygen in the Earth's crust, the chalkophilic elements having low abundances tend to concentrate in small amounts of concentrated sulphides, providing rare elements such as silver and bismuth in technically exploitable high concentrations. The elements in the crust represent a residue of about 0.4 percent compared with the abundances in the Solar atmosphere[21], and contrary to the cosmic distribution, the elements with Z above 30 have a roughly constant concentration of the order of magnitude 10^{-6} (1 g/ton) in the Earth, perhaps because of accumulated supernova dust. When condensing Solar material, H_2, He, and the main constituents CH_4 and NH_3 of the atmospheres of Jupiter and Saturn, leave at first. Since lithium, beryllium and boron very rapidly undergo thermonuclear reactions in the interior of stars, oxygen is the first common element in Terrestrian material, and actually almost half the mass of the crust. It is not too probable that the Earth has an interior mantle of sulphides of chalkophilic elements, unless the total amount of sulphur is unexpectedly large.

Returning to complex formation constants, J. Bjerrum[6] introduced a corrected constant taking the 55 M concentration of water into account. For unidentate ligands, he considers $\log K_n + 1.74$ or $\log K_{av} + 1.74$, the constant 1.74 being log 55. It is empirically known[22] that bidentate ligands such as ethylenediamine = 1,2-diamino-ethane $NH_2CH_2CH_2NH_2$ or multidentate ligands such as diethylenetriamine $HN(CH_2CH_2NH_2)_2$ or *tren* = *tris*(2-aminoethyl)amine $N(CH_2CH_2NH_2)_3$ show high formation constants, much higher than of unidentate NH_3 or CH_3NH_2. We do not here discuss this *chelate effect* in detail. A major correction for using a molar concentration scale rather than a unit of concentration comparable to the liquid amine[23] can be obtained[6] by evaluating the new constant

$$(\log K_n/q) + 1.74 \qquad (14)$$

in the case where it is almost certain that the n'th q-dentate ligand replaces q water molecules coordinated to the central atom. Bjerrum[6] pointed out that for a given central atom, the corrected constants in Eq. (14) for pyridine, ammonia and cyanide complexes have the ratios 0.6 : 1 : 1.6 though, as we discuss below, certain central atoms such as copper(I) and palladium(II) have slightly higher affinity to pyridine than indicated by this rule. On the whole, these three ligands, have the stronger bonding to chalkophilic elements, but not to the extreme extent of sulphide and iodide.

Schwarzenbach developed complexometric titrations with multidentate ligands such as nitrilotriacetate $N(CH_2CO_2)_3^{-3}$ and the potentially sexidentate (two N and four O) ethylenediaminetetra-acetate and potentially octadentate (three N and five O) diethylenetriaminepenta-acetate. Such synthetic amino-acids form strong complexes with more central atoms than ammonia, including the alkaline earths and the lanthanides. Nevertheless, Schwarzenbach[24,25] pointed out that it is important for the analytical chemist to recognize the trends of lithophilic elements preferentially coordinating with fluorine and oxygen and chalkophilic elements with phosphorus, sulphur and iodine. Seen from this point of view, nitrogen and chlorine are intermediate types of ligating atoms, though approaching the latter group. Two sulphur atoms on the same carbon or phosphorus atom frequently form chelates with four-membered rings, such as the bidentate dithiocarbamates $R_2NCS_2^-$, xanthates $ROCS_2^-$, dithiophosphates $(RO)_2PS_2^-$ and dithiophosphinates $R_2PS_2^-$[26,27] whereas carboxylic groups usually are not bidentate except in the equatorial plane of the uranyl complex $UO_2(O_2CCH_3)_3^-$ and in oligomers bridging two different central atoms, such as $OBe_4(O_2CCH_3)_6$ and $M_2(O_2CCH_3)_4, 2 H_2O$ with more or less pronounced bonding between the two M = Cr, Cu or Rh.

Ahrland, Chatt and Davies [28] published in 1958 the paper which has been the best recognized for the discussion of (A) and (B) character. These authors pointed out that all central atoms choose two among the 24 conceivable series of formation constants with halide anions in aqueous solution, and are either showing (A) character in the case $F \gg Cl > Br > I$ or (B) in the case $F < Cl < Br < I$. Copper(II) and cadmium(II) are marginally (B) and indium(III) (A). Ahrland, Chatt and Davies also performed a systematic comparison of ligands differing only by the replacement of an oxygen atom by a sulphur atom and found clear evidence for (A) central atoms preferring the former and (B) the latter version of the two ligands. When comparing two ligands differing in a nitrogen or a phosphorus atom carrying one lone-pair (though one tends to concentrate on primary amines RNH_2 but tertiary phosphines R_3P) the typical (B) elements show a predominant preference for phosphorus. It is as if certain border-line elements more readily show (A) behaviour toward amines, and copper(II) is readily reduced by phosphines like it is by iodide and by cyanide so its suspected (B) character is difficult to prove. However, if the preference for fluoride compared with iodide is taken as the primary definition of (A) character, iron(III), cobalt(II) and nickel(II) are definitely of (A) types though not as violently as aluminium(III) and magnesium(II) with comparable ionic radii. In the few cases where higher homologs of the ligands could be studied (mainly the neutral molecules R_2S, R_2Se and R_2Te or R_3P, R_3As and R_3Sb) it was not always the case that the formation constants increase monotonically with (B) central atoms [28].

It may be regretted that the words (A) and (B) character make one think about the short-period description of the Periodic Table containing 8 columns though the eighth column contains the triad elements. It is generally argued that the A elements Ca, Sr, Ba, Ra are different from the B elements Zn, Cd, Hg and the idea association immediately presents itself that the differences can be described as (A) and (B) character, as would be even more true for alkaline metals compared with the coinage metals (which Mendeleev originally arranged in tetrads in view of the oxidation numbers of copper(II) and gold(III) higher than the number of the column). If one calls all the elements from Na to Ar for A and from Cu to Kr for B, reserving the classification B for *post-transition group* elements, the agreement is fairly good. But it is always dangerous to have the same word for *almost* the same concept, and it would be difficult to classify both iron and platinum as (A) elements, or both as (B) elements, for that matter. In the sense of Goldschmidt, nickel, gallium and tin are on the border-line between lithophilic and chalkophilic elements and occur in both oxidic and sulphidic minerals. Further on, the ligands were not called (A) or (B) though it would have made sense to ascribe (A) behaviour to fluoride and (B) to iodide. These are among the reasons why the words "hard" and "soft" seem more appropriate, as discussed in the next chapter.

Ahrland[29,30] has revived the old interest in comparing ΔH with ΔG of complex formation. It has turned out that the strong complexation of aluminium(III) with fluoride forming $Al(H_2O)_{6-n} F_n^{+3-n}$ from $Al(H_2O)_6^{+3}$ and nF^- has negative ΔG (whereas the almost impossible formation of $Al(H_2O)_5Cl^{+2}$ has positive ΔG) *but positive* ΔH. Said in other words, the formation of the fluoro complexes is a spontaneous endothermic reaction occurring because ΔS of Eq. (1) is so large. The reason why the solution of $Al(H_2O)_5F^{+2}$ has so much lower entropy, is so much more disordered, than a solution of $Al(H_2O)_6^{+3}$ and F^- can be ascribed to the strong hydrogen bonding between F^- and a large number of water molecules. This structure disappears when F^- gets coordinated to a central atom, but at the same time, one has to supply heat to dissociate the hydrogen bonds. This rather unexpected situation (which already seemed paradoxical to Julius Thomsen) occurs in general by the reaction between (A) central atoms and fluoride or oxygen-containing ligands (such as SO_4^{-2} and carboxylates RCO_2^-) with exception of OH^- (where the reactions frequently are exothermic). On the other hands the reactions between (B) central atoms and ligands such as iodide or cyanide show comparable, negative ΔH and ΔG, that is ΔS is rather unimportant. It may be noted that the strong discrepancy between negative ΔG and positive ΔH in (A)cases occurs for *anionic* ligands. Neutral ligands, such as ammonia and ethylenediamines show comparable values[31] for ΔG and ΔH when forming complexes with nickel(II) and copper(II). A simple reaction such as the neutralization of H_{aq}^+ by OH^- in dilute aqueous solution has $\Delta G_0 = 19.1$ and ΔH_0 13.36 kcal/mole clearly favoured by increasing disorder of the pure water formed.

D. Hard and Soft Anti-bases and Bases

Ralph G. Pearson[32] suggested in 1963 to generalize the idea of (A) and (B) character to *hard* and *soft* behaviour of both Lewis acids (called *anti-bases* by Jannik Bjerrum[33]) and the ligating bases. This idea (frequently called SHAB = soft and hard acids and bases) has been reviewed[34] and Pearson has edited a reprint volume with comments[35] A symposium about this idea[36] was organized by R. F. Hudson in 1965 at the Cyanamid European Research Institute (closed 1968) in Cologny, the Republic of Geneva, and described in the 31. May 1965 issue of Chemical Engineering News.

There is no doubt that Pearson did not suggest something entirely new, as we have already seen, but he generalized the concepts derived from the mass-action law (formation constants and Brønsted acidity) in a variety of new ways. Many chemists have felt that hard-hard interactions are a new name for electrovalent (ionic) bonding and soft-soft interactions for covalent bonding[37]. This is also a part of the truth, but other aspects are far more sophisticated and deserve detailed discussion. Other chemists sharply criticize the short and colloquial words "hard" and "soft". In the writer's opinion, this reluctance to accept new words is similar to the resistance to use "weak" and "strong" acids in a way independent on their titration capability as the number of equivalents pro litre. It may be noted that these words were used by chemists for many years before pH was introduced in such a way that an acid with negative pK now is called strong, with pK around 5 weak and around 10 very weak.

The Pearson principle is dual:

Hard anti-bases react preferentially with hard bases.
Soft anti-bases react preferentially with soft bases.

It may be noted that both statements refer to an actual affinity and not only to a relative decrease of the affinity in one direction. Nobody can seriously argue, in spite of Eq. (13), that the reaction between $Hg(ClO_4)_2$ and $2\,NaCN$ is driven by the formation of $NaClO_4$ (which anyhow is ionically dissociated in solution) and not, as the $-\Delta H = 48.7$ kcal/mole shows, by formation of two Hg–C bonds. In view of the calorimetric results discussed by Ahrland[29,30] it is neither trivial that the hard-hard interactions correspond to an actual affinity though the reactions may be endothermic. One might have expected that any hetero-atomic molecule or polyatomic ion always would have a background of Coulombic attraction, called Madelung energy[4,38]. Actually, the scales of affinity are quite different for ions reacting in the gaseous phase[37] as well as in solvents of low dielectric constants, and the hard-hard interactions become stronger in these cases. Nevertheless, the chemistry simply can be different for two ions with the same oxidation state and almost the same ionic radii, *e.g.* sodium(I) and silver(I); rubidium(I) and thallium(I); calcium(II) and mercury(II); strontium(II) and lead(II); erbium(III) and thallium(III). However, the assumption of strongly positive Madelung contributions to the dissociation energy of hetero-atomic compounds does not apply the same way in aqueous solution, where the *aqua ions represent the standard state.* As seen in the last chapters, the numerical extent of hydration energy needs careful study, already because a minor uncertainty in these huge quantities can modify the formation constants expected. There is a large number of complexes which cannot be studied in aqueous solution because the ligands loose the competition with water. One should not be deluded to think that these problems are simple; when an aqueous solution of silver(I) fluoride and another of calcium(II) iodide react quantitatively in analogy to Eq. (10) precipitating AgI and CaF_2 it looks like a perfect confirmation of the Pearson principle, the hard Ca^{+2} selecting the hard F^- and the soft Ag^+ selecting the soft I^-. However, the complex chemistry surrounding AgI is complicated; an excess of Ag^+ dissolves the precipitate as Ag_3I^{+2} with a central iodide bound to three Ag^+ ligands, and an excess of I^- produces AgI_3^{-2} in closer analogy to PbI_3^- and HgI_4^{-2}. On the other hand, the supernatant solution of CaF_2 contains its small concentration of calcium as Ca^{+2} and not as fluoro complexes. This is not true for certain other, almost insoluble fluorides. The supernatant solution over LaF_3 contains some LaF^{+2} and ThF_4 quite a lot of ThF^{+3} and ThF_2^{+2} if some Th^{+4} is added. In all of these ionic complexes, as well as in the precipitates, the Pearson principle is satisfied; but it did not *a priori* specify the solubility of the precipitates, only a tendency toward the constitution of the *inorganic chromophore*[40] MX_N where N X-atoms surround the M atom.

Pearson[32] suggested that metallic elements and alloys *ipso facto* are soft. This statement about surface chemistry is rather unfamiliar to complex chemists, though it is beyond doubt that catalysts consisting of platinum or other noble metals, catalyzing reactions of the type Eq. (12), readily become ineffective, "poisoned", by adsorption of phosphines, arsines or sulphur-containing compounds. The water on iron surfaces[41] shows Brønsted acidity quite different from iron(II) or iron(III) aqua ions, and the corrosion process of iron can be shifted into different pH regions by applying an external potential. A corollary to Pearson's classification of metallic surfaces as soft is that organic molecules capable of adding to metallic surfaces, such as CO, C_6H_6, C_2H_4 and other olefins, all are soft.

We come here to a question which is not explicitly analyzed by Pearson, whether it is by necessity or an accident that zerovalent or negative oxidation numbers of transition elements are intrinsically soft. These cases are not generally discussed in connection with formation constants, but the clear-cut evidence for soft behaviour is of preparative nature, as discussed in the next chapter. The organometallic chemistry of such low oxidation states confirms the soft character of the ligands CO and C_6H_6 and indicates also soft properties of cyclopentadienide $C_5H_5^-$, tropylium $C_7H_7^+$ and cyclo-octatetraenide $C_8H_8^{-2}$ all having electronic configurations somewhat related to benzene[42]. The latter ligands can produce remarkably high coordination numbers N; the sandwich complexes $Cr(C_6H_6)_2$ and $Cr(C_6H_6)_2^+$ have $N = 12$ and the isoelectronic $Fe(C_5H_5)_2$ and $Fe(C_5H_5)_2^+$ $N = 10$; and the uranium(IV) compounds $U(C_5H_5)_3Cl$ and $U(C_8H_8)_2$ both have $N = 16$.

Pearson[32] argued that for a given element, the increase from low to high oxidation numbers corresponds to a monotonic development from soft to hard behaviour. Though this is a frequent tendency, the writer doubts that this rule is universally valid. Thus, thallium(III) is perceptibly softer than thallium(I). Tl(III) is also an instance of a high ionization energy of a soft central atom, though it has been argued that soft systems have low ionization energies (this would not apply to metallic surfaces on the whole being chemically softer when the work function is high). A subtler exception may be nickel(II) not being as soft as nickel(IV) occurring not only in NiF_6^{-2} but also in sulphur-containing complexes such as $Ni(S_2CN(C_4H_9)_2)_3^+$, in solid $NiAs_2$ and in complexes of bidentate phosphines and arsines. A perhaps even more striking case is the co-existence of bromide and dithiocarbamate in the copper(III) complex[43] $Br_2CuS_2CN(C_2H_5)_2$. Recently, the quadratic chromophore Cu(III)P$_4$ and square-pyramidal Cu(III)P$_4$Cl have been realized[44] with a bidentate phosphine. The oxidation was performed with concentrated nitric acid, which immediately would attack the free ligand to form phosphine-oxides. We are here in presence of a general problem: ligands of low electronegativity are particularly favourable for covalent bonding to a central atom of comparatively high electronegativity (one may for instance evaluate *optical electronegativities*[45] from electron transfer spectra) but at the same time, the ligand is reducing and can be oxidized by the central atom, frequently to a dimer having lost two electrons (I^- to I_2, RS^- to RSSR etc.). However, the stabilization of the soft-soft compound by covalent bonding may sometimes retard kinetically or prevent thermodynamically the redox reaction between the reducing ligand and the oxidizing central atom. As Jannik Bjerrum says, covalent complexes have known to stop the extent of electron transfer at the right moment.

The writer[36] analyzed the behaviour of manganese in the various oxidation states from Mn(−I) to Mn(VII) and suggested the modification of Pearson's rule that *the softness of central atoms decreases monotonically as a function of increasing fractional charge.*

There is clear evidence that the $3d^8$ Mn(–I) in Mn(CO)$_5^-$ is even softer than $3d^6$ Mn(I) in Mn(CO)$_5$X to be discussed in the next chapter. It is also clear that anionic carbonyl complexes are rather different from monomeric halide anions, except that Mn(CO)$_5^-$ also forms an oxidized dimer (OC)$_5$MnMn(CO)$_5$ to which it is problematic[4] to ascribe the *preponderant configuration* $3d^7$ and the concomitant oxidation state Mn(0). However, the hard character of manganese achieves a maximum at the numerous $3d^5$ Mn(II) compounds of which the majority are high-spin ($S = 5/2$) to be compared with the invariantly diamagnetic ($S = 0$) manganese(I) complexes. Disregarding redox reactions, the chemistry of Mn(II) is remarkably similar to magnesium(II), as can already be seen from the common etymology of the names of the two elements. Mn(III) is already less distinctly hard, as can be seen from the existence of MnCl$_5^{-2}$ and Mn(S$_2$CNR$_2$)$_3$. It is difficult to be explicit about Mn(IV) which is best known in MnF$_6^{-2}$ and in mixed oxides, though it seems less extremely hard than Mn(II). The higher oxidation states Mn(V), Mn(VI) and Mn(VII) are mainly known from tetrahedral oxo complexes[19] and represent cases like Cr(VI), Ru(VII), Ru(VIII) and Os(VIII) where oxide ligands are more capable than fluoride ligands to stabilize high oxidation numbers. In aqueous solution, other examples of strongly stabilized oxo complexes are UO$_2^{+2}$ and the vanadyl(IV) ion VO(H$_2$O)$_4^{+2}$ both lacking proton affinity in strong perchloric acid, as can also be-seen from the ultra-violet spectra of OsO$_4$ and the mixed nitrido complex NOsO$_3^-$ [19]. The very slight evidence for Mn(VII) not being very hard is that the green MnO$_3$F and MnO$_3$Cl show comparable synthetic stability.

A rationalization compatible with the maximum hardness at Mn(II) preceded by pronounced soft Mn(I) and by moderate hardness in Mn(III) and higher oxidation numbers would be that the fractional atomic charge of manganese nowhere is higher than in Mn(II), and the highest in the octahedral chromophore Mn(II)F$_6$ occurring in the rutile-type MnF$_2$ and the perovskites KMnF$_3$ and RbMnF$_3$ where it seems to be close to 1.9, and slightly lower in Mn(H$_2$O)$_6^{+2}$. This conclusion can be derived from the *nephelauxetic effect*, the decrease of the parameters of interelectronic repulsion compared with the corresponding gaseous ion, here Mn^{+2} [4,39,46] and from approximate M. O. calculations including the Madelung potential[4, 47]. Calculations neglecting the Madelung potential suggest fractional charges well below +1 in agreement with Pauling's electroneutrality principle, but this idea is almost incompatible with the visible spectra of transition-group complexes[48]. We do not here discuss the even more serious problems for the related hybridization theory[48, 49]. which seems irrelevant outside the $2p$ group elements. The physical mechanism behind the softness disappearing for high fractional charge of the central atom (besides an obvious emphasis on Coulombic effects) may be the higher density of low-lying excited states both of very low oxidation states (the energy of the partly filled shell approaching many empty orbitals) and of high oxidation states (showing electron transfer spectra at low wave-numbers because the loosest occupied M. O. of the ligands have comparable energy with the partly filled shell). It is perfectly clear that we are looking for a high density of low-lying *efficient* states because the many energy levels of $4f^q$ (especially for q between 2 and

12)[20, 50] do not modify the chemistry of the trivalent lanthanides compared with a smooth interpolation between La(III) and Lu(III). The question of efficient states is related to the (fairly remote) importance of electric dipolar polarizabilities derived from roughly additive molar refractivities to be discussed in subsequent chapters. It is perfectly clear that the relation of softness to fractional charges at best produces a convergent re-iteration (and at worst a vicious circle) because the more covalent compounds have lower fractional charges of the central atom[4, 39] and hence, the softness of the ligand influences the central atom to become softer. In the next chapter, we discuss the empirical evidence for this inorganic symbiosis.

It may be noted that Pearson uses the word "acid" according to Lewis. It is not perfectly clear whether the soft-soft interactions in $Mn(CO)_5^-$ has the $Mn(-I)$ central atom as Lewis acid and CO as base, or whether the back-bonding (to be discussed in detail in later chapters) to the ligand is so strong that $Mn(-I)$ is the base. Anyhow, a much more profound difficulty is that it is by no means certain that the coordination process really consists of exactly one pair of electrons from the base being donated to an empty orbital of the Lewis acid (anti-base). Though the chemical bonding in H_2, HeH^+ and H_3^+ undoubtedly is due to two electrons and in H_2^+ to one electron, it is not easy to justify the general belief in N pairs of electrons in N orbitals bonding N atoms to a central atom[48]. Thus, the NaCl-type oxides, nitrides and carbides such as MnO, NiO, CdO, LaN, LuN and HfC have $N = 6$ of both elements, but there are simply not more than four filled orbitals in the outer shells of the anion. This argument is even more striking in the non-metallic CaF_2-type beryllium(II) carbide Be_2C having $N = 8$ for carbon. In the writer's opinion, there is no particular reason to look for six bonding orbitals in AlF_6^{-3}, SiF_6^{-2}, PF_6^- and SF_6 and hence, the classification of the central atoms as anti-bases (as would also be true for SiF_4 and PF_5) does not imply the absolute validity of the Lewis model, which has recently been discussed[52].

J. Bjerrum[53] has written a review about hard and soft anti-bases mainly applying ΔG arguments for selected systems. However, he has also brought a very interesting question up for debate. In the Brønsted theory, it is very clear-cut to talk about an acid as the proton adduct of its *corresponding base*. Now, one may define a base corresponding to a given anti-base as the adduct of an electron pair to the anti-base. Frequently, both species are known in chemistry. The simplest case is the proton H^+ as anti-base and H^- (to be discussed in the next chapter) as base. Another case is the anti-base I^+ and the base I^- where iodine(I) complexes are known. For instance elemental iodine disproportionates in pyridine under the influence of silver(I) nitrate in the metathetical reaction

$$I_2 + Ag(NC_5H_5)_2^+ = AgI + I(NC_5H_5)_2^+ \tag{15}$$

Another simple example is the soft anti-base Tl^{+3} adding two electrons to Tl^+ which is hardly a Lewis base though square-pyramidal green low-spin ($S = 1/2$) $Co(CN)_5^{-3}$ reacts[54] with aqueous Tl^+ to give $(NC)_5CoTlCo(CN)_5^{-5}$, where the distribution of oxidation states is debatable.

In the case of polyatomic entities, the stereochemistry changes frequently drastically by adding two electrons to an anti-base. For instance, monomeric SO_3

is planar like BF_3 but the corresponding base SO_3^{-2} is pyramidal like IO_3^- also containing a lone-pair. One may observe even more profound changes. Thus, diborane $H_2BH_2BH_2$ is a well-known "electron-deficient" molecule though the writer considers it as hydride bridged and isosteric with Al_2Cl_6 also consisting of two tetrahedra joined along an edge. There is nothing unusual about hydride bridges; hafnium(IV) boranate $Hf(H_3BH)_4$ has $N = 12$ of the central atom in the molecule, and the perovskite $BaLiH_3$ has each barium(II) surrounded by 12 hydrides, lithium(I) six (like in crystalline LiH) and each hydrogen($-$I) is octahedrally coordinated with two lithium on one Cartesian axis and four barium on the two other. Anyhow, adding two electrons to B_2H_6 modifies the stereochemistry to the anion $B_2H_6^{-2}$ isosteric with ethane and having a central B$-$B bond. It can also be argued that $Mn(CO)_5X$ contain the square-pyramidal $Mn(CO)_5^+$ anti-base re-arranging to trigonal-bipyramidal $Mn(CO)_5^-$ when adding two electrons.

Already Pearson[32] noted the differences in softness, normally increasing by adding an electron pair, as in the carbonium ion $C(C_6H_5)_3^+$ not being as soft as the carbanion $C(C_6H_5)_3^-$ or, *a fortiori*, aliphatic examples such as $C(CH_3)_3^+$ and $C(CH_3)_3^-$. The writer is only aware of one case of a less soft base, *viz.* Tl^+, being formed by adding two electrons to a soft anti-base. Bjerrum[53] proposes that the anti-base corresponding to a soft base is hard (of which many examples are known, such as SO_3) but also that the anti-base corresponding to a hard base is soft, in analogy to the definite result in Brønsted theory that a weak base corresponds to a strong acid (formed by addition of a proton) and that a strong base corresponds to a weak acid. It may be worthwhile noting that this statement does not discuss whether a weak base itself is an acid; it may be a strong acid like HCl, a very weak acid like H_2O, or not being able to be a Brønsted acid at all (not containing protons) like Cl^-.

It is by no means certain that the anti-base corresponding to a hard base always is soft. Bjerrum[53] assumes that F^+ is a soft anti-base. This is very difficult to disprove because fluorine(I) is not known in compounds, not even in the type Eq. (15). OF_2 is better classified[4] as an oxygen(II) fluoride, and the hypofluorite molecules such as CF_3OF (trifluoromethanol cannot be made from the deprotonated $Rb^+CF_3O^-$ because it decomposes to HF and OCF_2), F_5SOF and O_3ClOF all involve oxygen rather than fluorine in anomalous high oxidation number. In a Gedankenexperiment, F^+ has a higher affinity to I^- forming IF than to F^- forming F_2 along the lines of Eq. (13) but at the same time, there is no doubt that F^+ would oxidize hard compounds such as CeF_3 and TbF_3. Anyhow, a distinct exception to this opinion is UF_6 and UF_6^{-2} where UF_6 is only known to react as a hard anti-base with F^- forming adducts such as UF_8^{-2} whereas the uranium(IV) complex formed by adding two electrons to UF_6 also is hard.

In this connection, it may be remembered that Brønsted acids sometimes form "partly neutralized" adducts with their corresponding bases, such as FHF^- and $H_2OHOH_2^+$ known from crystal structures. If H_{aq}^+ on an instantaneous picture is $H_9O_4^+$ this would also fall in this category. It is less frequently realized[52] that an anti-base may react with less than the maximum amount of base. Thus, the anti-base XeF^+ and its adduct with XeF_2, symmetrical (but bent) $FXeFXeF^+$ and fluoride-bridged Siamese-twin octahedra $F_5SbFSbF_5^-$ and $F_5PtFPtF_5^-$ are all known from recent crystal structures. Originally, Bjerrum[33] called SbF_6^- an anti-base + base (anba) adduct, SbF_5 being the anti-base and F^- the base. However, for practical purposes, SbF_6^- is also a base, potentially able to bridge another atom in a complex, as in $Sb_2F_{11}^-$ or in the hypothetical $F_5SbFBF_3^-$. When gaseous PCl_5 crystallizes as $PCl_4^+PCl_6^-$ it shows that the same molecule can be a base and an anti-base to a much more pronounced extent than water is simultaneously a base and a Brønsted acid. To the writer, an anba adduct remains a base because its potential capability of reacting with another anti-base releasing base. In the case where the anba adduct can add further base, it is at the same time an anti-base. Hence, one tends to ascribe "hard" or "soft" character to an anba adduct without specifying whether it is considered a base or anti-base. SiF_6^{-2} is uniformly hard and PtI_6^{-2} soft all over.

E. Inorganic Symbiosis and Preparative Chemistry

It would be closing the eyes to half of the content of Pearson's classification to consider only systems where complex formation constants have (or can) been measured. If one wants to stabilize unusually low oxidation numbers, it is well-known that the best chances are to select the ligands among the class H^-, I^-, R_3P, CO, CNR, CN^-, ... that is Pearson's soft bases. Actually, a very large number of complexes containing triphenylphosphine $L = P(C_6H_5)_3$ have been reported the last years, and it is also known[55] that PF_3 is particularly successful in stabilizing negative oxidation states such as $Ru(PF_3)_4^{-2}$ and $Ir(PF_3)_4^-$. What is important to the synthetic chemist is that a mixture of soft ligands frequently yields stable complexes. Among the homogeneous catalysts for hydrogenation of olefins, Wilkinson's rhodium(I) compound RhL_3Cl and Vaska's iridium(I) compound $Ir(CO)L_2Cl$ are able to form a lot of interesting adducts. Thus, the latter complex is oxidized by H_2 to the iridium(III) hydrido complex $IrH_2(CO)L_2Cl$ whereas the adducts of oxygen are known in certain cases[4] suggesting octahedral iridium(III) with a peroxo (O_2^{-2}) ligand and others having iridium(I) bound to O_2 much in the same way as ethylene in the platinum(II) complex $PtCl_3(C_2H_4)^-$ discovered by Zeise in 1829. Vaska's compound has an isomer[51] where Ir(III) has extracted a hydride from a phenyl group forming also an Ir–C bond Malatesta prepared Ir(III) IrH_3L_3. It must be added in all fairness that the iridium(V) compound IrH_5L_2 also is known, the only other well-characterized case of Ir(V) being IrF_6^-. The tendency toward high N and high oxidation number in hydrido complexes can also be seen in the rhenium(VII) complex ReH_9^{-2}.

It is worthwhile to analyze why co-existing soft ligands assist low oxidation numbers. If we want to make a copper(I) compound, it is very difficult to try the aqua ion, the fluoride or the anhydrous sulphate because they disproportionate to the metallic element and a higher oxidation state, here Cu(II). However, as seen in Eq. (7) it is easier to make the ammonia complex $Cu(NH_3)_2^+$ under anaerobic conditions, and even easier to make copper(I) complexes of pyridine and of conjugated bidentate ligands such as 2,2'-dipyridyl and 1.10-phenanthroline. The experimental problems are reversed in the case of iodides and cyanides, where it is easy to precipitate CuI or CuCN or to prepare solutions in an excess of the ligand containing CuI_2^-,

$Cu(CN)_2^-$ or $Cu(CN)_4^{-3}$ but a difficult kinetic problem to detect the intermediate Cu(II) complexes.

Manganese(I) is known in a few compounds with five CO ligands and a sixth ligand, H^-, CH_3^-, Cl^-, Br^- or I^-. It is possible to have only four CO ligands in the dimeric $(OC)_4MnBr_2Mn(CO)_4$ having two bromide bridges analogous to niobium(V) chloride crystallizing as dimeric octahedra $Cl_4NbCl_2NbCl_4$. The soft cyclopentadienide provides a compound with only three CO as the molecule $C_5H_5Mn(CO)_3$. Said in other words, one needs a lot of soft ligands to keep manganese(I). Until now, $Mn(CO)_5F$ has not been prepared. One has to be very cautious about saying that it cannot conceivably exist; many compounds included in the fascinating book "Non-existent Compounds"[56] have later been prepared, such as $ClFCl^+$, ClF_6^+, BrO_4^- and a rich folklore of xenon compounds. However, it is almost certain that $Mn(CO)_5F$ would disproportionate:

$$4\ Mn(CO)_5F = (OC)_5MnMn(CO)_5 + 2\ MnF_2 \qquad (16)$$

driven by the stability of manganese(II) fluoride. Under anhydrous conditions, it is possible to prepare salts of $Mn(CO)_6^+$. They undergo a most unexpected reaction in aqueous solution

$$Mn(CO)_6^+ + H_2O = Mn(CO)_5H + CO_2 + H_{aq}^+ \qquad (17)$$

where one of the CO ligands is oxidized to CO_2 and one hydrogen(I) from water simultaneously reduced to a hydride ligand. This does not prevent that $Mn(CO)_5H$ is a Brønsted acid in aqueous solution with pK = 7.1 undergoing a redox reaction with the hydride ligand:

$$Mn(CO)_5H = Mn(CO)_5^- + H_{aq}^+ \qquad (18)$$

This acid-base reaction would be felt by many chemists to show that the ligand is not exactly hydrogen(−I). However, arguments can be given[4] that all transition-group to hydrogen bonds involve H(−I). Thuse, $Co(CO)_4H$ is trigonal-bipyramidal compatible with 3 d^8 Co(I) isoelectronic with $Fe(CO)_5$ and $Mn(CO)_5^-$ but nevertheless a strong acid readily forming $Co(CO)_4^-$ in aqueous solution, which is a tetrahedral 3 d^{10} system like $Fe(CO)_4^{-2}$ and $Ni(CO)_4$. $PtL_2(CO)H$ is quadratic (5 d^8) and not tetrahedral, etc.

It is one of the virtues of Pearson's classification that the behaviour of hydride as a super-soft ligand has been clarified, though it is outside the domain of complex formation constants. Like the σ-bonded carbanions of the type CH_3^-, $(CH_3)_3CCH_2^-$, $C_6H_5CH_2^-$ and $(CH_3)_3SiCH_2^-$ it shows many of the effects of *trans*-influence discussed below, but H^- cannot be suspected of conventional back-bonding nor hyperconjugation. Nevertheless, the constituents of water pose rather specific problems. It has already been mentioned that OH^- is not a typically hard base. It is possible in the first approximation[40] to ascribe the tendency of water coordinated to the element M in the oxidation state z to deprotonate to OH^- (and then further to O^{-2}) to an increasing value of z divided by the ionic radius. Thus, M(VII) and M(VIII) are almost exclusively coordinated by O^{-2} in aqueous solution. In sufficiently acidic solution, S(VI) forms mixed hydroxo-oxo complexes such as $HOSO_3^-$ and Cr(VI) and Te(VI) more readily so with their larger radii of the central atoms. Only the largest M(IV) such as Th^{+4} and U^{+4} are known as aqua ions, whereas the typical behaviour of M(III) is to change from aqua to hydroxo ligands at a pH which increases in the series Fe(III)<Al(III)~Cr(III)~Rh(III)<Sc(III)<Lu(III)<....<La(III). The so-called diagonal similarities in the Periodic Table can be understood as the influence of the ionic radius. Be(II) is very similar to Al(III) and finishes as a pure hydroxo complex $Be(OH)_4^{-2}$ by its amphoteric behaviour above pH = 12, a behaviour shown by no other alkaline earth. On the other hand, $B(OH)_3$ does not react at low pH (though concentrated sulphuric acid forms tetrahedral $B(OSO_3H)_4^-$) but deprotonates to oxo complexes in strongly alkaline solution like silicic acid, etc.

However, this simple description is not appropriate for certain oxidizing central atoms forming strong covalent bonds. Already $Fe(H_2O)_6^{+3}$ is far more acidic than expected (but $Fe(OH)_3$ is so insoluble that it is not particularly amphoteric) and Cu(II) is too acidic for a M(II) of that size. Extreme cases are $Pd(H_2O)_4^{+2}$ having the same pK as U_{aq}^{+4} and the univalent halogens comparable to P(V). It is perhaps significant that hypochlorite OCl^- does not seem to hydrate to $Cl(OH)_2^-$ in aqueous solution. It may very well be that pK, of ClOH relative to the latter species would have been more normal. Anyhow, it is perfectly clear that one cannot divide elements M in a given oxidation state z into "chemical metals" and "metalloids" according to whether $M(OH)_z$ is a base or an acid. The development is perfectly continuous.

The word *inorganic symbiosis*[57] was proposed in connection with a striking effect of the other ligands bound to cobalt(III). It must be admitted that both Co(II) and Co(III) are border-line cases with a mild preference for fluoride like indium(III)- though the chemistry is very different in other respects because of "ligand field" stabilization[4,42,58] apparently increasing the formation constants of complexes of ammonia, amines, N-bound NO_2^- and cyanide producing large sub-shell energy differences Δ[4,50] between the two anti-bonding 3d-like orbitals with angular functions proportional to (x^2-y^2) and $(3z^2-r^2)$ some 3 to 4 eV above the three other 3d-like orbitals. If the groundstate keeps the same S (here zero) and it is assumed that the parameters of interelectronic repulsion are the same (the vary in practice according to the nephelauxetic series) the "ligand field" stabilization is proportional to the difference of Δ_c in the complex and Δ_a in the aqua ion, and is $- 1.2 (\Delta_c - \Delta_a)$ for octahedral $d^3 (S = 3/2)$ and $d^8 (S = 1)$ and $- 2.4(\Delta_c - \Delta_a)$ for $d^6 (S = 0)$ systems. Since halides have lower Δ than aqua ions, this may explain a discrimination against chloride in chromium(III) and nickel(II) not getting beyond $Cr(H_2O)_4Cl_2^+$ and $Ni(H_2O)_5Cl^+$ in 12 M HCl to be compared with perceptible amounts of $Mn(H_2O)_4Cl_2$ and $ZnCl_4^{-2}$. Since Δ for fluoride falls in the *spectrochemical series* between chloride and water, this trend might also explain a mildly harder behaviour of nickel(II) than of zinc(II).

The four complexes $Co(NH_3)_5X^{+2}$ are kinetically robust, but the ΔG values are in favour of X = F and the subsequent X = Cl, Br and I are monotonically weaker bound. This hard behaviour, one may argue, is superposed an unusually high affinity for ammonia making the chemistry of Co(III) quite different[58] from Sc(III), Fe(III) and Ga(III) though in neutral solution, Eq. (6) is followed by other reactions finally *precipitating* dark brown $Co(OH)_3$. The "ligand field" stabilization $- 2.4 (\Delta_c - \Delta_a) =$

-11300 cm^{-1} $= -32.2$ kcal/mole corresponds to a contribution 23.7 to log β_6 which would have been 11.5 only without this effect of the partly filled shell. The "purpureo" complexes Co(NH$_3$)$_5$X^{+2} serve as convenient starting materials for replacing X by another ligand. Adamson found to his great surprise that the reaction with CN$^-$ proceeds along a quite unexpected route

$$Co(NH_3)_5 X^{+2} + 5\,CN^- = Co(CN)_5 X^{-3} + 5\,NH_3 \qquad (19)$$

which he explained involved the kinetic intermediate Co(CN)$_5^{-3}$ formed by traces of Co(II). Yellow Co(NH$_3$)$_5$CN^{+2} was first prepared much later[59]. Anyhow, the thermodynamic properties of Co(CN)$_5$X^{-3} clearly show *soft* characteristics by having X = I stronger bound than Br. Another analogy to Mn(CO)$_5$X is that hydride and benzyl (and other alkyl) derivatives readily can be made by oxidation of Co(CN)$_5^{-3}$ with H$_2$ and by splitting of alkyl iodides in analogy to the synthesis of Grignard reagents:

$$2\,Co(CN)_5^{-3} + H_2 = 2\,Co(CN)_5 H^{-3} \qquad (20)$$
$$2\,Co(CN)_5^{-3} + C_6H_5CH_2 I = Co(CN)_5 CH_2 C_6 H_5^{-3} + Co(CN)_5 I^{-3}$$

Other cobalt(III) complexes showing soft behaviour with preference for cyanide, iodide and alkyl ligands are the cobalamine group of vitamin B 12[60] and the bis (dimethylglyoximates) involving four nitrogen atoms bound in the equatorial plane in both cases, as is true for porphyrin and chlorophyll complexes.

One may attempt to explain the soft character of the anti-base Co(CN)$_5^{-2}$ in various ways. The strong nephelauxetic effect[4,46] in Co(CN)$_6^{-3}$ compared with Co(NH$_3$)$_6^{+3}$ indicates a lower fractional charge of the central atom in the former case, as also shown by careful X-ray diffraction studies of the electron density[61] in [Co(NH$_3$)$_6$] [Co(CN)$_6$] and one may return to the first interpretation of Pearson's dual principle, that the decreased charge of the cobalt atom makes it softer. However, an alternative explanation involves back-bonding transferring electron density from the d-shell to the empty orbitals of CN$^-$ (or CO) having a node between the two atoms of the ligand. It is clear that such a change may increase the covalent bonding of iodide, hydride etc. but it is less obvious that it does not encourage electrovalent bonding to fluoride. It may be that the second alternative only serves to make the fractional charge of central atoms with negative oxidation number less negative. Anyhow, the importance of symbiotic effects for preparative chemistry cannot be explained away. For comparison with Eq. (17) it may be noted that Malatesta[62] prepared colourless manganese(I) complexes of isonitriles CNR such as Mn(CNC$_6$H$_5$)$_6^+$ which can be oxizidized with strong nitric acid to dark violet Mn(CNC$_6$H$_5$)$_6^{+2}$. The corresponding E^0 above + 1 V clearly shows a contrast with harder ligands usually bound to Mn(II). From a practical point of view, the worst difficulty is to make complexes of soft central atoms with hard ligands. The opposite problem of hard central atoms with soft ligands can be solved by direct reaction in vacuo or in noble gases between the metallic element (say, barium or lanthanides) and elemental selenium, tellurium or iodine.

One might have expected that an obvious application of the idea of inorganic symbiosis would be *ambidentate ligands* such as thiocyanate SCN$^-$. Crystal structures show that typically soft central atoms are bound to the sulphur end, such as $Cu(SCN)_2^-$, $Rh(SCN)_6^{-3}$, $Pd(SCN)_4^{-2}$, $Ag(SCN)_2^-$, $Au(SCN)_4^-$, $Hg(SCN)_2$ and $Hg(SCN)_4Co$ whereas most other central atoms are bound to the nitrogen end, such as $Cr(NCS)_6^{-3}$, $Fe(NCS)_6^{-3}$, $Mo(NCS)_6^{-3}$ and, interestingly enough, $Nb(NCS)_6^-$ and $Ta(NCS)_6^-$. From reflection spectra of crystalline salts, it can be concluded[63] that octahedral $Co(NCS)_6^{-4}$ and $Ni(NCS)_6^{-4}$ occur, whereas organic solvents may contain tetrahedral $Co(NCS)_4^{-2}$ and $Ni(NCS)_4^{-2}$. Whereas $Cr(H_2O)_5NCS^{+2}$ is the stable species, it is possible[64] to prepare $Cr(H_2O)_5SCN^{+2}$ (re-arranging within an hour) by reacting Cr_{aq}^{+2} with $Co(NH_3)_5NCS^{+2}$. Ragnar Larsson pointed out that certain species such as $Cd(SCN)_4^{-2}$ seem to re-arrange to a considerable extent in solution (according to infra-red spectra) and it cannot be excluded that $Rh(NCS)_x(SCN)_{6-x}^{-3}$ occur. It is beyond doubt that mixed complexes of palladium(II) of amines or phosphines exist with N-bound thiocyanate. However, it is a question whether the N-end is so much harder than the S-end. It has two lone-pairs (like N_3^-) and not one lone-pair like an amine, the electron transfer spectra[45, 63] of N-bound isomers show an optical electronegativity 2.8 comparable to bromide, and the spectrochemical position is close to water. Interestingly enough, protonation to species such as $Co(NH_3)_5NCSH^+$ and external addition, of Ag^+ or Hg^{+2} to the S-ends[65] as first detected by Werner[1] produces spectra similar to $Co(NH_3)_6^{+3}$ showing that the ligands NCSH, NCSAg and $NCSHg^+$ have Δ equal to ammonia and probably have exactly one lone-pair available. The co-existence of tetrahedral $Co(II)N_4$ and $Hg(II)S_4$ in solid $Co(NCS)_4Hg$ is another instance of thiocyanate bridging two atoms of two metallic elements. Other ambidentate ligands such as selenocyanate have been much discussed[63, 66] but like the old example of N- and O-bound nitrite NO_2^- the conclusions about hard and soft behaviour have not been clear-cut. More interesting cases might occur if fluorine-containing organic compounds might form bridges to lanthanides or thorium. It is of great importance for biochemistry that sulphur-containing amino-acids incorporated in proteins have a specific affinity for Fe(III), Cu(II) and Zn(II), and in the case of intoxications, Hg(II) and Pb(II).

II. Spontaneous Deviations from the Highest Symmetry Available

A. Absorption Band Intensities and Stereochemistry of Copper(II)

It is generally argued that electrovalent bonding does not show strongly *preferred bond angles*. Whereas cubic and hexagonal close packing of identical spherical atoms both have $N = 12$ (slightly below 4π) it is rare for ionic binary salts to show N higher than 8, the value found for CsCl, though $LaCl_3$ has $N = 9$ and both $SrTiO_3$ and K_2PtCl_6 have $N = 12$ for the large ion (Sr^{+2} and K^+). The disadvantage of not showing the highest possible symmetry is not extreme on a Madelung picture[4] but in practice, ionic crystals seem to minimize their electrostatic energy. It must be remembered that the angular dependence of covalent bonding may not be as pronounced as frequently assumed, as seen from the existence of P_4 and cyclopropane, and in particular that the hybridization model cannot exclusively explain the bond angles[48, 67, 68]. Nevertheless, the low $N = 4$ of diamond, silicon, grey tin and GaAs indicate covalent bonding, and $N = 2$ known from the isoelectronic complexes $Au(NH_3)_2^+$, $Hg(NH_3)_2^{+2}$, $Hg(CH_3)_2$ and $Tl(CH_3)_2^+$ cannot be explained with geometrical arguments based on relative ionic radii where $N = 4$ or 6 rather would be expected, as known from crystalline $[Co(NH_3)_6]TlCl_6$ to be compared with thallium(III) in strong hydrochloric acid, where only $TlCl_4^-$ is detected.

It is very instructive to compare high-spin ($S = 1$) nickel(II) with copper(II) bound to oxygen- and nitrogen-containing ligands [23, 69]. The octahedral chromophore $Ni(II)O_6$ is found in the aqua ion in solution and in many salts, in complexes of many organic ligands [70], in $Ni_xMg_{1-x}O$ and many mixed oxides [71], in glasses [72] and in definite compounds such as $NiCO_3$ (discrete carbonate anions), $NiTiO_3$ (ilmenite with $N = 6$ for titanium) and $Ba_2[Ni(OH)_6]$. The absorption bands in transparence or in reflection spectra are exceedingly characteristic for octahedral Ni(II). The fundamental conclusion [4] is that we are in presence of *eight d*-like electrons in the preponderant electron configuration. However, for our purposes, the band intensities are informative and show striking contrasts with copper(II). To the writer's knowledge, it has not been pointed out that moderate deviations from the highest symmetry available to the chromophore, producing comparatively high intensities of Laporte-forbidden transitions and a variety of observable effects, are characteristic for soft central atoms. At an instantaneous picture, $Cu(H_2O)_6^{+2}$ and other instances of $Cu(II)O_6$ are *not* cubic, *i.e.* having three equivalent Cartesian axes.

We are not here emphasizing the Jahn-Teller effect conserving the center of inversion, if originally present, and related to the holohedric part of the "ligand field" [67]. It is well-known that the numerical extent of the first-order Jahn-Teller distortion is only large in the systems having two or three anti-bonding orbitals occupied in an *unbalanced fashion* in the high symmetry (by 0 and 1; or by 1 and 2 electrons) such as octahedral high-spin ($S = 2$) d^4 exemplified by Cr(II) and Mn(III) and d^9 mainly known from Cu(II). The Jahn-Teller effect is weak in $Ti(H_2O)_6^{+3}$ and VCl_4 and hardly perceptible in lanthanide compounds.

Seen from the point of view of complex formation constants [6] there is no doubt that the Jahn-Teller effect contributes to $N_{char} = 4$ in the copper(II) amine complexes whereas an early surprise was that $Ni(NH_3)_n(H_2O)_{6-n}^{+2}$ shows no evidence for this characteristic coordination number. However, the effect in closer connection with our subject is that $N_{max} = 5$ in $Cu(NH_3)_5^{+2}$ as is also true for the mixed complex with ethylenediamine $Cu\,en_2(NH_3)^{+2}$ and recently [73] the $Cu\,en_3^{+2}$ known to have $N = 6$ in the crystalline sulphate (and a reflection spectrum rather similar [74] to tris-complexes of 2,2'-dipyridyl and 1,10-phenanthroline) has been shown to have $N = 5$ in aqueous ethylenediamine with two bidentate and one unidentate ligand. The solutions of $[Cu\,en_2](ClO_4)_2$ and of $[Cu\,en_2]Cl_2$ in aqueous ethanol [75] have absorption spectra which are all linear combinations of two spectra, one belonging to the purple aqueous species (which may be $Cu\,en_2(H_2O)^{+2}$) and another a blue-violet species with the constitution $Cu\,en_2(C_2H_5OH)^{+2}$ or $Cu\,en_2(ClO_4)^+$. It has been demonstrated [76] that *anhydrous* $Cu(NH_3)_4^{+2}$ incorporated in the p-toluenesulphonate $[Pt(NH_3)_4](CH_3C_6H_4SO_3)_2$ is pink and quite different from the blue aqueous form, which most probably has $N = 5$ and is square-pyramidal $Cu(NH_3)_4(H_2O)^{+2}$. It is important to realize that the fact that $K_4 = 120$ but $K_5 = 0.3$ in the copper(II) ammonia complexes does not by itself prove that the so-called perpendicular fifth ligand is not strongly bound. It is conceivable that all five ligands in $Cu(NH_3)_5^{+2}$ are bound with comparable distances and force constants, but have elongated the four Cu—N distances in the equatorial plane to such an extent that the free energy is less negative than in $Cu(NH_3)_4(H_2O)^{+2}$. No crystal structure has been shown to contain $Cu(NH_3)_6^{+2}$ [77].

The *oscillator strength P* (called *f* by many authors) is proportional to the area of an absorption band, the molar extinction coefficient ϵ as a function of the wave-number σ in cm^{-1} (called ν in many cases). When the band is a Gaussian error curve $\epsilon = \epsilon_0\, 2^{-(\sigma - \sigma_0)^2/\delta^2}$ with the maximum ϵ_0 at σ_0 and the one-sided half-width δ, the area is $2.1289\, \epsilon_0\delta$ and

$$P = 4.32 \cdot 10^{-9} \int \epsilon d\sigma = 9.20 \cdot 10^{-9}\, \epsilon_0\delta \qquad (21)$$

Whereas the absorption band in the red of the methylene blue cation has P close to 1, it is very rare that bands of inorganic complexes in the visible (13000 to 25000 cm^{-1}) have P above 0.1. The electron transfer band of PtI_6^{-2} with maximum at 20250 cm^{-1} has $P = 0.18$ whereas the electron transfer band giving the well-known purple colour of MnO_4^- is a vibrational structure in the green having $P = 0.03$. In a way, it is surprising that P-values in the ultra-violet covered by most spectrophotometers (down to 192 nm or up to 52000 cm^{-1} or 6.5 eV) rarely add up to more than a few-tenths because it is a quantum-mechanical *sum rule* that the total P of all electric dipole transitions of a system containing q electrons is exactly q. However, nearly all of the oscillator strength corresponds to transition at higher energy in the far ultra-violet and X-ray region, as discussed below in the chapter on refractive indices.

In chromophores possessing a center of inversion, not only the one-electron wave-functions ψ (the orbitals) but also the total wave-functions Ψ choose between *even* and *odd parity* (like even or odd *l* in the spherical symmetry of a monatomic entity)[67]. It is a necessary condition for a transition to be allowed as electric dipole radiation to take place between states of opposite parity. This *selection rule* is very carefully obeyed in atoms (the rare exceptions are due to weak electric quadrupole[a] or magnetic dipole transitions, or to perturbation by adjacent atoms at high pressures of the gas emitting the spectral lines) but experience shows that both gaseous molecules[78] and compounds containing partly filled shells in condensed matter[69] show *Laporte-forbidden* transitions between a groundstate and an excited state both having even parity (or odd in the case of an odd number of *f* electrons). It has been recognized for many years[79] that the origin of the non-vanishing oscillator strength P of such transitions in polyatomic systems is that the nuclei do not remain stationary at the positions defining the point-group of high symmetry (such as O_h of a regular octahedron) but deviate slightly on an instantaneous picture[67] due to

---

a) Lanthanide(III) compounds show a few *hypersensitive transitions*[207] showing selection rules like quadrupole transitions, but they are not due to static distortions but are highly dependent on the ligating atoms, such as conjugated β-diketonates[208] and strongest in gaseous iodides[209]. Though the intensities can be described with Judd-Ofelt parameters[210] this phenomenon is closely related to chemical softness and is perhaps due to a rapidly varying local dielectric constant.

their vibration (even at 0 °K). What has been much more slowly recognized [69, 80] is that this *vibronic coupling* of even states mainly occur through excited states of odd parity corresponding to electron transfer bands at the wave-number σ_a. Actually, one can determine an empirical formula depending on the wave-number σ_f of the forbidden transition considered, and the oscillator strength P_a of the electron transfer band:

$$P_f = P_a H_{af}^2/(\sigma_a - \sigma_f)^2 \tag{22}$$

where the effective non-diagonal element H_{af} turns out to be between 1500 and 2500 cm^{-1} of the same order of magnitude as the full half-width 2δ of the absorption bands. Hence, if the distance $(\sigma_a - \sigma_f)$ is ten times larger, and P_a about 0.1, Eq. (22) predicts P_f close to 10^{-3}.

A list of representative P values has been published[81] and a few cases are given in Table 1. It is seen that the sum of P_f of the three spin-allowed (but Laporte-forbidden) transitions of Ni(H$_2$O)$_6^{+2}$ is only $1.3 \cdot 10^{-4}$ and in Ni(NH$_3$)$_6^{+2}$ $2.3 \cdot 10^{-4}$. These values are quite characteristic for M(II) of the $3d$ group with exception of Cu(II). If the chromophore lacks a center of inversion, one can induce transition dipole moments without the help of Eq. (22) and tetrahedral CoX$_4^{-2}$ have P close to 0.01. However, this mechanism is remarkably ineffective in approximately octahedral complexes (such as *orthoaxial chromophores* [67, 82]) having all the ligating atoms on the Cartesian axes) and M(NH$_3$)$_5$X^{+2} (M = Cr, Co, Rh, Ir) mostly have P comparable with the corresponding M(NH$_3$)$_6^{+3}$. Actually, since δ is somewhat larger, ϵ_0 of Eq. (21) is lower for X = Cl.

From this point of view, the rapidly increasing P from $3.5 \cdot 10^{-4}$ for the combined shoulder and band of the copper(II) aqua ion to $2.3 \cdot 10^{-3}$ for Cu(NH$_3$)$_5^{+2}$ having a spectrum of the same form cannot be explained exclusively from the denominator in Eq. (22). Thus, the first electron transfer bands are situated approximately at 50000 cm^{-1} in the Cu(II) aqua ion and at 60000 cm^{-1} in Ni(H$_2$O)$_6^{+2}$. Even considering the fact that several transitions coincide in the copper(II) complexes (it is recognized[76, 77] that the shoulder is due to the excitation of one electron from $(3z^2 - r^2)$ to the available empty position in $(x^2 - y^2)$ and the maximum to excitation of (xz, yz) degenerate in systems with two equivalent Cartesian axes x and y, and of (xy) to this hole in $(x^2 - y^2)$ whereas species such as Cu(NH$_3$)$_4$(H$_2$O)$^{+2}$ apparently presenting one, somewhat asymmetric band have all these transitions almost coinciding) it would be difficult to explain a factor higher than 2 as the intensity ratio between Cu(II) and Ni(II) with the same ligands. Certain multidentate amines acieve quite impressive P values as seen in Table 1.

Energy-wise (discussing the positions σ_f of the internal transitions in the partly filled shell) it has been known since long time[6, 23] that the variation is far less regular in Cu(II) than in Ni(II). As long nickel(II) complexes remain high-spin $(S = 1)$ the mixed complexes of the type Ni(NH$_3$)$_n$(H$_2$O)$_{6-n}^{+2}$ have band positions closely following the *rule of average environment*[83] as if they were hexakis-complexes of a hypothetical ligand having $\Delta = (n\Delta_c + (6 - n)\Delta_a)/6$. The development of the spectra adding from 1 to 4 ammonia ligands to Cu(II) is also linear though σ_c of the tetrammine is 1.34 times larger than σ_a of the aqua ion to be compared

Table 1. Wave-numbers σ (in the unit 1000 cm^{-1}) and oscillator strengths P (in the unit 10^{-5}) of spin-allowed (but Laporte-forbidden) transitions in nickel(II), copper(II) and palladium(II) complexes of water, ammonia, ethylenediamine (en), diethylenetriamine (den) and pyridine (py). Shoulders in parentheses

	σ	P		σ	P		σ	P
Ni(H$_2$O)$_6^{+2}$	8.5	3	Cu(H$_2$O)$_x^{+2}$	(9.4)	10	Pd(H$_2$O)$_4^{+2}$	(24.4)	80
	13.8, 15.2	3		12.6	25		26.4	200
Ni(NH$_3$)$_6^{+2}$	25.3	7	Cu(NH$_3$)$_4$(H$_2$O)$^{+2}$	16.9	120	Pd(NH$_3$)(H$_2$O)$_3^{+2}$	27.8	320
	10.75	5.3	Cu(NH$_3$)$_5^{+2}$	(11.8)	60	Pd(NH$_3$)$_2$(H$_2$O)$_2^{+2}$	29.3	580
	17.5	7		15.6	170	Pd(NH$_3$)$_4^{+2}$	33.9	470
Ni en$_3^{+2}$	28.2	10	Cu en$_2$(H$_2$O)$^{+2}$	18.2	140	Pd en(H$_2$O)$_2^{+2}$	29.4	570
	11.2, (12.4)	10	Cu en$_3^{+2}$	(11.8)	~100	Pd en$_2^{+2}$	35.0	800
	18.35	10		16.4	~200	Pd den(H$_2$O)$^{+2}$	31.6	1300
Ni den$_2^{+2}$	29.0	14	Cu en$_2$(NH$_3$)$^{+2}$	(13)	–	Pd den(OH)$^+$	32.1	1100
	11.5, (12.4)	20		16.7	~200	Pd denCl$^+$	29.8	1200
	18.7	12	Cu en$_2$py^{+2}	16.8	~250	Pd den$_2^{+2}$	33.9	~800
Ni py$_4$(H$_2$O)$_2^{+2}$	29.1	19	Cu den(H$_2$O)$_x^{+2}$	16.3	170	Pd den(NH$_3$)$^{+2}$	33.6	1200
	10.15	6	Cu den$_2^{+2}$	(11.8)	120	Pd(den)py^{+2}	33.5	1400
	16.5	8		15.9	250	Pd py$_4^{+2}$	36.4	~700
Ni en$_2^{+2}$ ($S=0$)	27.0	15	Cu den(NH$_3$)$^{+2}$	17.4	200	Pd(OH)$_4^{-2}$ (?)	27.2	~400
	22.2	~100	Cu py$_4^{+2}$	17.0	~150			

with $(\Delta_c/\Delta_a) = 1.25$ in regular octahedral complexes [4, 69]. The pentammine effect[5] is that this ratio *regresses* to 1.24 in $Cu(NH_3)_5^{+2}$. The *tetragonality ratio* σ_{Cu}/σ_{Ni} known[69] to vary between 1.1 for *tris*-complexes of dip = 2,2'-dipyridyl and phen = 1,10-phenanthroline to 1.73 for the *bis*(ethylenediamine) complex (with the aqua ion 1.48) is undoubtedly an increasing function of the deviation from cubic octa-hedral to quadratic coordination around Cu(II). It is difficult to know exactly what the tetragonality ratio would be for a regular octahedron though an extrapolation from $V(H_2O)_6^{+2}$ to $Ni(H_2O)_6^{+2}$ suggests a value close to 0.9.

It is worthwhile once more to make a distinction between the tendency toward tetragonality (long and weak bonds to one or two ligands on the axis perpendicular to the plane containing four strongly bound ligands at short distance) which can be interpreted as first-order Jahn-Teller distortion (Reinen[84] has made a careful study of the mixed oxides of many types) and the spontaneous loss of the centre of inver-sion, as when $Cu(NH_3)_5^{+2}$ does not bind a sixth ligand. Quite general, the stereo-chemistry of copper(II) is rather complicated[85] and whereas *cis*-Cudip$_2(H_2O)_2^{+2}$ and *cis*-Cuphen$_2(H_2O)_2^{+2}$ are strikingly different from *trans*-complexes of aliphatic amines and the mixed[86] Cu(en)dip^{+2}, one also knows trigonal-bipyramidal Cuphen$_2I^+$. Like the high-spin trigonal-bipyramidal chromophores $M(II)N_4X$ formed[87] by the quadridentate $N(CH_2CH_2N(CH_3)_2)_3$ the intensities are not exceedingly high, where-as low-spin $M(II)P_4X$ and $M(II)As_4X$ formed by sterically hindered quadridentate phosphines and arsines[88] have $\epsilon_0 \sim 5000$ and $P \sim 0.1$ of transitions which *energy-wise* can be described by "ligand field" theory, and in particular, the angular overlap model. The latter cases are even more extreme than CoX_4^{-2}. Actually, the orange $CuCl_4^{-2}$ and purple $CuBr_4^{-2}$ belong to the point-group D_{2d} and are intermediate between the regular tetrahedron (which can be selected as four of the eight corners of a cube) and a square by being four of the corners of a parallele piped with two identically long and one short side. However, it appears that $[Pt(NH_3)_4][CuCl_4]$ contains planar complexes. On the other hand[89] trigonal-bipyramidal $CuCl_5^{-3}$ is also known. It must in all fairness be admitted that nickel(II) is not universally octa-hedral, but can exhibit other stereochemistry with soft ligands. Besides quadratic chromophores, diamagnetic ($S = 0$) nickel(II) also form square-pyramidal $Ni(II)As_4X$ and $Ni(CN)_5^{-3}$ and one example of high-spin square-pyramidal $Ni(II)O_5$ is known. Certain crystal structures contain square-pyramidal $InCl_5^{-2}$ and (Jahn-Teller-unstable) $MnCl_5^{-2}$.

It is not easy to explain Irving and Williams' rule that complex formation con-stants show a maximum for Cu(II) between Ni(II) and Zn(II) as a consequence of "ligand field" stabilization. It is true that multidentate ligands sterically predisposed to low ratios of tetragonality (such as 1.35 for ethylenediaminetetra-acetate) also show this effect to a less pronounced extent. Thus, $\log K_1$ can be compared[6, 22, 23] with bidentate "en" and tridentate "den"

	Mn(II)	Fe(II)	Co(II)	Ni(II)	Cu(II)	Zn(II)	
NH_3	–	–	2.1	2.8	4.1	2.4	
en	2.7	4.3	5.9	7.6	10.7	5.9	
den	4.0	6.2	8.1	10.7	16.0	8.9	(23)
enta^{-4}	14.0	14.3	16.2	18.6	18.8	16.3	

However, softer ligands show a much greater propensity to bind Cu(II) though the measurements sometimes are rendered impossible because of redox reactions. It is particularly interesting that Dunitz and Orgel[90] compare heats of formation of binary compounds from gaseous ions of the six elements of Eq. (23). In this case, we are not comparing a "ligand field" stabilization of a complex with the aqua ion, but a more fundamental $-\Delta H$. For the solid fluorides, the variation is roughly like the last line of Eq. (23) but the solid chlorides show a pronounced maximum at Cu(II). This is also true for the oxides, but the comparison is more difficult, because MnO, FeO, CoO and NiO crystallize in NaCl type, ZnO has $N = 4$ in BeO type, and CuO is its own type with quadratic coordination. The same kind of criticism can be directed toward the sulphides, selenides and tellurides, but there is no doubt that the Cu(II) compounds are far more stable than Ni(II) and Zn(II).

B. Palladium(II) Complexes

Only a minority of nickel(II) complexes are low-spin with the $(x^2 - y^2)$ orbital empty whereas high-spin palladium(II) is exceedingly rare. Two instances are solid PdF_2 and $CsPdF_3$. It is possible to study a variety of ligands bound to Pd(II) in solution, and it was pointed out[69] that a ratio σ_{Pd}/σ_{Rh} closely similar to the tetragonality ratio σ_{Cu}/σ_{Ni} can be defined as the ratio between the wave-numbers of the first spin-allowed transition in the Pd(II) and in the rhodium(III) complex with the same ligands. This ratio varies between 1.02 and 1.10 (with exception of the amphoteric solution in 2 M NaOH possibly containing $Rh(OH)_6^{-3}$ and $Pd(OH)_4^{-2}$ where the ratio is 1.136 to be compared with 1.037 for the aqua ions). The approximate invariance of this ratio is due to a compensation of the two effects that quadratic $Pd(II)X_4$ have unusually short Pd–X distances (actually, $PdCl_4^{-2}$ and $PdCl_6^{-2}$ have the same Pd–Cl distance in spite of the differing oxidation state) and that the sub-shell differences normally are larger in M(III) than in M(II). It can be argued that the first spin-allowed transition in $Pd(II)X_4$ and $Rh(III)X_6$ correspond essentially to the excitation of an electron from (xy) to $(x^2 - y^2)$ changing the electronic density in the equatorial plane.

When looking through the catalog "Stability Constants"[3] one finds hundreds of ligands for which the formation constants have been determined for a definite class of bivalent central atoms consisting of the six elements from Eq. (23), Mg(II), Zn(II), Cd(II), Hg(II) and Pb(II). With exception of the last, these are also the ammonia complexes of M(II) studied by Bjerrum[6]. The reason for this emphasis on eleven bivalent elements is that one readily can provide the aqua ions in perchlorate solution, whereas noble metals [with exception of Ag(I)] most frequently are furnished as chloro complexes such as $RhCl_6^{-3}$, $PdCl_4^{-2}$, $IrCl_6^{-3}$, $IrCl_6^{-2}$, $PtCl_4^{-2}$, $PtCl_6^{-2}$ and $AuCl_4^{-}$. Actually, metallic palladium is a limiting case of a noble metal by being soluble in hot nitric acid like silver. However, the brown solution obtained contains nitrato complexes with higher ϵ than the yellow aqua ion which has a quite characteristic spectrum with a shoulder at 24 400 cm^{-1} and a maximum with $\epsilon_0 = 79$ at 26 400 cm^{-1}. This species was first obtained in solution[69] by adding an excess of $Hg(ClO_4)_2$ in 1 M $HClO_4$ to $PdCl_4^{-2}$. Since $\log K_1 = 4.5$ for $Pd(H_2O)_3Cl^+$ is considerably smaller than $\log K_1 = 6.7$ for $HgCl^+$, the chloride is almost quantitatively taken over by the mercury(II). On the other hand, it is *not* possible to obtain the aqua ion by dissolving the hydroxide in perchloric acid. The coffee-brown solution obtained is partly colloidal, change with time and has far too high ϵ suggesting polymeric hydroxo complexes. The same tendency can be seen[91] in the ultra-violet spectrum of $Rh(OH)_3$ dissolved in strong $HClO_4$ which only forms pure $Rh(H_2O)_6^{+3}$ after boiling for a considerable time.

As seen in Table 1, most Pd(II) complexes have spectra remarkably similar in shape to copper(II) complexes of the same ligands, but with twice as high wavenumbers. It cannot be absolutely excluded that the yellow aqua ion (exceptionally having a shoulder 2000 cm^{-1} before the maximum) is $Pd(H_2O)_5^{+2}$ with a weakly bound fifth water ligand. This would also explain why the colourless amphoteric species has higher σ (which is unheard about among hydroxo complexes) if it is quadratic $Pd(OH)_4^{-2}$. A related phenomenon is that $Pd(NH_3)_4^{+2}$ increases ϵ_0 from 200 to 210 going from a solution in dilute ammonia to 1 M $NaClO_4$ (92,93). This marginal effect might be ascribed to the formation of square-pyramidal $Pd(NH_3)_4(OClO_3)^+$ with unidentate perchlorate weakly bound perpendicular to the plane containing $Pd(II)N_4$. On the other hand, no penta-coordinated complexes can be detected in the ultra-violet spectra of $Pd(NH_3)_4^{+2}$ in 12 M NH_3 nor in $Pd(CN)_4^{-2}$ in 1 M KCN. Quite generally, the tendency to go from quadratic ($N = 4$) to square-pyramidal ($N = 5$) coordination is weaker in Pd(II) than in low-spin Ni(II) and much weaker than in Cu(II).

However, it would appear that the spontaneous refusal to retain a centre of inversion in palladium(II) is connected with the equatorial plane containing the four ligating atoms rather than by out-of-plane distortions. Both Lene Rasmussen and the writer were students of Jannik Bjerrum in 1950 when we started to discuss how one may determine formation constants of Pd(II), and when she returned from the period 1964–67 as professor at the Ife University in Ibadan, Nigeria, and participated in a UNESCO conference in Geneva, we took up this problem again, which curiously enough had not been solved in the meantime. Our first step was to show that anhydrous or weakly hydrated $PdSO_4$ (which can be made from boiling mixtures of nitric and sulphuric acid with powdered Pd) dissolves in 1 M $HClO_4$ with the same spectrum as the aqua ion, and is not changed if SO_4^{-2} is removed with barium perchlorate. A 0.01 M $PdSO_4$ in 0.1 M $HClO_4$ has the same spectrum (excepting a weak increase above 40000 cm^{-1}) but does not keep more than a few days without beginning decomposition to the coffee-like products. Our next contribution was to prepare $[Pd(NH_3)_4](CH_3C_6H_4SO_3)_2$ because it is imperative to establish equilibrium from both sides in acidic solutions containing known quantities of NH_4^+ and where pH is measured for use in Eq. (5). The formation constants found by us[92] for ammonia and ethylenediamine, by Elding[94] for chloride and bromide, and by Chang and Bjerrum[95] for the phosphine $P(C_6H_5)_2(C_6H_4SO_3^-)$ = dpm$^-$ are:

	$\log K_1$	$\log K_2$	$\log K_3$	$\log K_4$	
NH_3	9.6	8.9	7.5	6.8	
en	>21	18.4			(24)
Cl^-	4.47	3.29	2.41	1.37	
Br^-	5.17	4.25	3.30	2.22	
dpm$^-$	10.2	9.8	6.3	4.9	

One of the most remarkable results is that K_2 is so much larger than K_3 for the ammonia and phosphine complexes, where K_2/K_3 is 25 and 3100 respectively, to be compared with the statistical value 2.25 for $N = 4$. Said in other words, the two first unidentate ligands are strongest bound, as if there is a tendency toward the point-group C_{2v}. Correspondingly, there are good reasons to believe that $Pd(NH_3)_2(H_2O)_2^{+2}$ predominantly occurs as the *cis*-isomer. There has been many studies of the kinetics of Pd(II) also involving mixed amine-chloro complexes[93,96-100]. The results summarized in Eq. (24) clearly show the soft character of Pd(II) as central atom. This can also be seen from the study by Srivastava and Newman[101] of the spectra of the mixed complexes $PdCl_{4-n}I_n^{-2}$ where the equilibrium constant for the exchange of the first chloride with iodide is 9000, for the second iodide 12 000, for the third 600 and the replacement with the last iodide 20. It is not easy to study the intermediate $Pd(H_2O)_{4-n}I_n^{-2}$ because black PdI_2 is so insoluble, but $\log \beta_4$ for PdI_4^{-2} must be 12.15 larger than 11.54 for $PdCl_4^{-2}$ in Eq. (24) or 23.7. That this value is smaller than $\log \beta_4 = 32.8$ for $Pd(NH_3)_4^{+2}$ can be explained with "ligand field" stabilization[92].

The reason why $\log K_1$ is almost impossible to measure of $Pd \, en(H_2O)_2^{+2}$ is that with $pK_1 = 7.5$ of enH_2^{+2} and $pK_2 = 10.2$ of enH^+, the free ethylenediamine concentration [en] is $10^{-17.7}$ times the concentration of protonated ethylenediamine at pH = 0. Hence, H_{aq}^+ cannot perceptibly decompose the mono-complex (without assistance of other ligands such as Cl^-) in 10^{-2} M concentration at positive pH (which is almost a condition for reasonable conditions of activity) if $\log K_1$ is above 21. This argument is much stronger for diethylenetriamine, where $pK_1 = 4.7$, $pK_2 = 9.2$ and $pK_3 = 10.0$ of $denH_3^{+3}$. If $\log K_1$ is above 26, it would be difficult to determine. A very interesting aspect of the reaction between $Pd(H_2O)_4^{+2}$ and $denH_3^{+3}$ in 1M $HClO_4$ is that an intermediate is formed[102] with a spectrum similar to $Pd \, en(H_2O)_2^{+2}$ and most probably *cis*-$Pd(II)N_2O_2$ with two of the nitrogen atoms of the ligand bound to the central atom, and the third presumably protonated. Bidentate diethylenetriamine also occurs in $Pd \, den_2^{+2}$ having a spectrum like $Pd \, en_2^{+2}$ without any evidence for $N = 5$ (as would be true for $Cu \, den_2^{+2}$) nor $N = 6$. However, the intermediate is transformed in 1M $HClO_4$ at room temperature to the same species as one obtains by removing Cl^- from $Pd \, denCl^+$ with Ag^+ or Hg^{+2}. The simplest proposal for the constitution of this species is $Pd \, den(H_2O)^{+2}$ but this is not perfectly certain in view of somewhat unexpected complications[102] of the deprotonation to $Pd \, den(OH)^+$. The spectra of such species have been reported[103] and may be compared with the species $PdLX^+$ [104,105] formed by the ligand N,N,N'',N''-tetra(ethyl)diethylenetriamine $HN(CH_2CH_2N(C_2H_5)_2)_2$ previously used by Basolo as a sterically hindered tridentate amine for kinetic studies.

Turning from chemical to spectroscopic properties, the ϵ_0 values and the oscillator strengths increase dramatically from the aqua ion ($P = 0.003$ which already is 20 times larger than the sum of the spin-allowed, but Laporte-forbidden, transitions in $Ni(H_2O)_6^{+2}$) toward the amine complexes in a way strongly dependent on the branching of the multidentate ligands. The observation that the band intensities of $Pd \, en(H_2O)_2^{+2}$ are lower than of $Pd \, en_2^{+2}$ indicates spontaneous distortions of, at least, the latter species. It may be noted that the electron transfer band of Pd(II) aqua ions is close[92,93] to 50000 cm^{-1} like Cu(II) and hence, the high intensity cannot be explained by Eq. (22). For comparison it may be mentioned that the two spin-allowed transitions at $19\,300$ and $24\,300$ cm^{-1} of RhCl$_6^{-3}$ have $P = 1.3 \cdot 10^{-3}$ each whereas the first electron transfer band at $39\,200$ cm^{-1} has $P = 0.8$. Comparison with isoelectronic species such as PdCl$_6^{-2}$ suggests a second electron band close to $52\,000$ cm^{-1} with P about 2. It is not excluded that Rh(III) also shows spontaneous deviations from a centre of inversion.

The situation is already more complicated with diethylenetriamine-den and the tetra-ethyl-substituted L. It is quite striking that the tridentate[102] Pd den(H$_2$O)$^{+2}$ has $\epsilon_0 = 520$ higher than $\epsilon_0 = 420$ of Pd den(OH)$^+$ and $\epsilon_0 = 460$ of Pd den Cl$^+$ as if the mixed anion complexes were closer to be planar, as seems also to be the case for Pd den(NH$_3$)$^{+2}$ having $\epsilon_0 = 486$[106] and a band position at $33\,700$ cm^{-1} slightly below other Pd(II)N$_4$ in Table 1. The corresponding PdL(H$_2$O)$^{+2}$ and PdLX$^+$ have ϵ_0 between 1140 (aqua ion) and 700[104,105]. For comparison, it may be mentioned that most diethylenetriamine complexes such as Ni den$_2^{+2}$, Co den$_2^{+3}$ and Rh den$_2^{+3}$ have band positions closely similar to the tris(ethylenediamine) complexes and intensities somewhat higher, whereas the spectrochemical position is lower[107] in Mn den$_2^{+2}$ than in Mn en$_3^{+2}$, perhaps for some steric reason.

Livingstone[108] prepared a variety of Pd dip X$_2$ and Pd phen X$_2$ containing the chromophore cis-Pd(II)N$_2$X$_2$. Though the first bidentate ligand is much stronger bound than the second, it is not trivial that Livingstone also prepared Pd dip$_2^{+2}$ and Pd phen$_2^{+2}$ because M phen$_2$ X$_2^+$ (M = Cr, Co, Rh, Ir) are only known[109] as cis-isomers, and it is usually argued that Van der Waals repulsion between hydrogen atoms of the two ligands destabilize the co-planar coordination in the trans-isomer. Sigel[110] has also argued that Cu phen$_2$(H$_2$O)$_2^{+2}$ is cis because of definite effects of π-backbonding, but it is remembered that Cu(en)dip^{+2} [86] is roughly quadratic. Anyhow, crystal structures[111] containing Pd dip$_2^{+2}$ show the point-group D_{2d} of Pd(II)N$_4$ (like CuCl$_4^{-2}$) diminuishing the repulsion between the ligands, and Pd phen$_2^{+2}$ seems also to be distorted[112] from planarity. Lene Rasmussen and the writer[113] studied the two residual positions of the anti-base Pd phen^{+2} in complexes such as Pd phen(NH$_3$)$_2^{+2}$, Pd(phen)en^{+2} and Pd(phen)py$_2^{+2}$. In connection with the latter pyridine complex, we also showed by preliminary measurements that log K_3 of Pd py$_3$(H$_2$O)$^{+2}$ is 6.5 (the same value is found for the binding of pyridine in Pd(den)py^{+2}).

Hence, the corrected value from Eq. (14) is 8.24, i.e. 89 percent of 9.24 for K_3 of the ammonia complexes in Eq. (24) constituting an exception from J. Bjerrum's rule[6] of this ratio being 60 percent. It would appear that very soft central atoms favour pyridine, as suggested by preparative evidence of the relative ease of making the chromium(O) complexes Cr(CO)$_5$py and Cr(CO)$_5$NH$_3$.

Lene Rasmussen and the writer[113] also prepared a moderately soluble, bright yellow, crystalline Pd phen(OH)$_2$ which slowly transforms in an orange, amorphous wax possibly containing oxide-bridged polymers. The crystals are soluble in an excess of OH$^-$ to a species which seems to be five-coordinated Pd phen(OH)$_3^-$ with a rather different ultra-violet spectrum according to Parthasarathy[114]. The Bronsted acidity of mixed aqua-phenanthroline complexes is not as high as of Pd(H$_2$O)$_4^{+2}$ but markedly higher than of water in mixed complexes of aliphatic amines where the pK values scatter between 6 and 8[102,114].

Kasahara[115] prepared a carbanion ligand L^- by deprotonating 2-phenylpyridine on the *ortho*-position on the phenyl ring in lemon-yellow $LPdCl_2PdL$ with two chloride bridges, each chromophore being $Pd(II)CNCl_2$. L^- is isoelectronic with 2,2'-dipyridyl. The formation of anion bridges is a frequent consequence of preparing neutral complexes in non-polar solvents, and Schurter[116] and the writer have started an investigation of monomeric species such as $LPd(H_2O)_2^+$. Unfortunately, all known salts of this cation are almost insoluble, as it is also true[114] for $Pd\,phen(H_2O)_2^{+2}$. Further on, unidentate nitrate or perchlorate tend to replace aqua ligands in these instances of $Pd(II)CNO_2$. N,N-dialkylbenzylamines[117] also react with $PdCl_4^{-2}$ coordinating with the nitrogen atom and the deprotonated carbon atom in *ortho*-position of ligands such as $C_6H_5CH_2N(C_2H_5)_2$ forming a five-membered ring with the palladium atom. Parshall[118] has written an excellent review of such *orthopalladation reactions* which Trofimenko[119] prefers to call cyclopalladation. The same benzene ring can bind two palladium, as in derivatives of N,N,N',N'-tetra-alkyl-*p*-xylene-$\alpha\alpha'$diamines of the type $C_6H_4(CH_2N(C_2H_5)_2)_2$ deprotonating on two carbons on one side of the ring. Deprotonated phenylhydrazine L^- also forms $LPdX_2PdL$ and $LPdX_2^-$[120]. Other instances of activated carbon sites are the formation of aniline from Cu(II) or Pd(II) benzoate and ammonia at 200 °C studied by Arzoumanidis and Rauch[121]. A case of orthoiridation by hydride abstraction from coordinated triphenylphosphine was recently reported[51].

Besides the question of carbanions being formed in the presence of palladium(II) with concomitant chelation, the chemistry of the anti-bases LPd^+ are clear-cut examples of the *trans*-influence, the ligands *trans* to strongly bound, soft ligands (such as carbanions and hydride) have long distances and weak bonding. It is proposed by Venanzi to reserve the word *trans*-effect to the kinetic variations which can be quite dramatic because of the large change of rate constants for a comparatively small change of the activation energy in the Arrhenius equation. A typical *trans*-influence is the highly increased pK values of water *trans* to carbanions. On the whole, quadratic gold(III) complexes are even more acidic than Pt(IV). For instance, $Au(H_2O)Cl_3$ is a strong acid forming H_{aq}^+ and $AuCl_3(OH)^-$ whereas pK of $PtCl_5(H_2O)^-$ is close to 4. Harris and Tobias[122] studied the methyl complexes *cis*-$Au(CH_3)_2(H_2O)_2^+$ and *fac*-$Pt(CH_3)_3(H_2O)_3^+$ having an aqua ligand *trans* to each carbanion ligand with the moderate pK_1 values 8 and 7, respectively. In a certain sense, these complexes are extreme cases of an evolution seen in Eq. (24) where the softest ligands have the largest K_2/K_3. The tendency toward $N_{char} = 2$ (being bent in a right angle like H_2Te and not linear such as $HgCl_2$) may correspond to an intrinsic tendency toward C_{2v} on the instantaneous picture of a species such as $Pd(NH_3)_4^{+2}$ explaining the high intensities first pointed out by Lene Rasmussen and the writer[102]. One may suggest that the inherent deviations from a centre of inversion are less pronounced in Pt(II) than Pd(II). Both $PtCl_4^{-2}$ and $IrCl_6^{-3}$ have weaker bands than the isologous $PdCl_4^{-2}$ and $RhCl_6^{-3}$. This might be explained as higher wave-numbers σ_a of electron transfer bands, under equal circumstances, in the $5d$ than in the $4d$ group, but since σ_f increase almost as much, Eq. (22) is not a perfectly convincing argument. The situation is quite different in gold(III) complexes generally having σ_a and σ_f coinciding[123] and the kinetics is far more rapid of Au(III) than of the isoelectronic Pt(II). One reason may be that five-coordinated intermediates are easier to obtain in the former case.

Another reason may be electron transfer catalysis like in the Co(II) and Co(III) couple. Since $Pt(H_2O)_4^{+2}$ is almost certain to disproportionate to metallic platinum and Pt(IV) (though Elding[124] has separated $Pt(H_2O)_3Cl^+$ and *trans-* and *cis-* $Pt(H_2O)_2Cl_2$) complex formation constants are usually given relative to $AuCl_4^-$ and $PtCl_4^{-2}$. It is generally expected that Pt(II) is softer than Pd(II) but there is very little quantitative evidence available. Chang and Bjerrum[95] find $\log K_1$ and $\log K_2$ for *cis-*$Pt(NH_3)_2dmp_2$ 11.5 and 11.1 and for the *trans-*isomers 11.5 and 10.5, slightly above 10.2 and 9.8 for $Pd(dmp)_n(H_2O)_{4-n}^{+2-n}$.

In addition to the encyclopedia of Gmelin and of Pascal, Griffith[125] has written a comprehensive book about the chemistry of ruthenium, rhodium, osmium and iridium, and Hartley[126] two volumes about palladium and platinum with emphasis on metallo-organic chemistry.

C. Mercury(II)-Complexes

The colourless Hg_{aq}^{+2} existing in perchloric acid (in which HgO can be dissolved) is almost as acidic as Pd(II) with pK close to 2. It is known to bind two unidentate ligands[3,6] very strongly:

	$\log K_1$	$\log K_2$	$\log K_3$	$\log K_4$	$\log \beta_4$	
NH_3	8.8	8.7	1.0	0.8	19.3	
Cl^-	6.7	6.5	1.0	1.0	15.2	
Br^-	9.0	7.9	2.3	1.7	20.9	(25)
I^-	12.9	10.9	3.7	2.3	29.8	
CN^-	18.0	16.7	3.8	3.0	41.5	
dmp^-	14.5	10.2	5.1	2.6	32.4	

(to be compared with $\log \beta_4 = 22.0$ for $Hg(SCN)_4^{-2}$ and $\log \beta_2 = 23.4$ for $Hg\,en_2^{+2}$) with exception of fluoride, where solid HgF_2 (crystalling in CaF_2 type with $N = 8$) dissolves mainly as the aqua ion with $\log K_1$ at most 1. The characteristic coordination number $N_{char} = 2$ corresponds to a linear coordination preferred by d^{10} systems in low, but positive, oxidation numbers such as Cu(I), Ag(I), Au(I). Whereas usual thallium(III) complexes are tetrahedral ($N = 4$) or octahedral ($N = 6$), metallo-organic species such as $Tl(CH_3)_2^+$ contain a linear CTlC group.

The spectroscopic properties of d^{10} systems are less informative than of complexes containing partly filled shells. Tetrahedral HgX_4^{-2} have bands in the ultraviolet[127] intermediate in character between $5d \rightarrow 6s$ excitation of the central atom and electron transfer from the filled orbitals of X^- to the empty $6s$ orbital, whereas the bands[45, 140] of thallium(III) and lead(IV) halides have the latter origin. It is now recognized[58] that the ionization energies of the d-shell in Zn(II), Cd(II) and Hg(II) between 14 and 18 eV are considerably larger than the values found for Co(III), Cu(II), Pd(II) and Au(III) compounds by photo-electron spectra. The situation is different in gaseous $Ni(CO)_4$ (8.8 and 9.7 eV), $Ni(PF_3)_4$ (9.55 and 10.58 eV), solid CuCN (9 eV) and silver(I) compounds (10 to 11 eV).

The silver(I) aqua ion has definite $4d \rightarrow 5s$ transitions at 44700, 47500 and 51900 cm^{-1} [128] but the cadmium(II) and mercury(II) aqua ions have no characteristic bands before vacuo ultra-violet, where thin layers of water also are opaque. This is one reason why N is not known. A plausible instantaneous picture is octahedral $Hg(H_2O)_6^{+2}$ distorted in the opposite direction of $Cu(H_2O)_6^{+2}$ with two short and four long Hg–O distances. As far one can know, zinc(II) and cadmium(II) aqua ions are regularly octahedral.

It is clear that first-order "ligand field" stabilization does not apply to closed-shell d^{10} complexes. Nevertheless, Pouradier and the writer[129] found the regularity that $\log \beta_2$ for two ligands L and L' differ 1.84 times as much for gold(I) as for silver(I). The corresponding ratio of variation of $\log \beta_2$ for copper(I) and silver(I) is close to 1.25 but is less constant. Though this approach has been criticized[130] it still constitutes an unexpected relation between gold(I) and silver(I) chemistry. It may be noted that the factor 1.84 is independent of the standard oxidation potential of metallic gold E^0 to the hypothetical Au_{aq}^+ (which has been estimated between +1.6 V and +2.1 V by different authors). We rationalize this behaviour by a second-order "ligand field" stabilization of linear XMX by admixture of the rotationally symmetric d orbital $(3z^2 - r^2)$ with the empty s orbital, increasing the electronic density in the equatorial plane and rendering the central atom oblate, as previously suggested by the writer[131] and Orgel[132]. By the way, the detailed distribution of excited levels of $PtCl_4^{-2}$ can best be understood[133] as an admixture of $5d$ and $6s$ with the opposite sign of the linear combination depleting the electronic density of $(3z^2 - r^2)$ in the equatorial plane.

In the absence of "ligand field" stabilization of the partly filled shell favouring amine and phosphine complexes (with sub-shell energy differences larger than in the aqua ion) and discriminating against halide complexes, Eq. (25) suggests that Hg(II) rather is softer than Pd(II) in Eq. (24). Actually, the generic name "mercaptanes" for RSH come from the strong bonding of RS^- to Hg(II), and certain phosphines are even stronger bound to Hg(II) than the mono-sulphonate of triphenylphosphine dmp^-. Thus, Meier[134] find $\log \beta_2 = 37.3$ and $\log K_3 = 5.2$ for $(C_2H_5)_2PCH_2CH_2OH$ which is more soluble in water than triethylphosphine.

The cyanide complexes of Hg(II) are somewhat paradoxical by having so large formation constants and, at the same time, reacting very rapidly. The main reason seems to be the re-arrangement from $N = 2$ to 4. Slowly reacting Hg–C bonds are known from the numerous organo-mercury compounds HgR_2 with aliphatic ligands as in $Hg(CH_3)_2$ or aromatic, such as $Hg(C_6H_5)_2$. It has attracted interest recent years that HgR^+ undergo many reactions, also contributing to the metabolism and toxicity of mercury in organisms. The equilibrium constant for the reaction

$$Hg^{+2} + C_6H_6 = HgC_6H_5^+ + H_{aq}^+ \tag{26}$$

has been determined[135] to be close to 300, unfortunately in nitric acid, which is slightly suspected of forming additional nitrato complexes.

The addition of one or two other ligands to $HgCH_3^+$ has been studied by Schwarzenbach and Schellenberg[136]. Though the idea of inorganic symbiosis would suggest that this cation may be considered as the typical soft anti-base, it is not excluded that the carbanion ligand has a certain *trans*-influence modifying the soft behaviour. Among the many curious reactions of this cation is that sulphide forms $S(HgCH_3)_3^+$ in analogy to sulphonium ions SR_3^+.

The *mercuration* of organic compounds can go very far with a lot of C—H replaced by C—HgX groups. As a curiosity may be mentioned $C(HgI)_4$ containing the lowest carbon content (0.9 percent) known in any organic compound.

Nyholm and Vrieze[137] prepared rhodium(III) complexes of the ligand HgX^- together with three R_3As and two X^-, by replacing a hydride ligand with HgX_2 eliminating HX. In a certain sense, this ligand exemplifies the oxidation state mercury(0) isoelectronic with thallium(I). It may be stretching the analogy a little too far to argue that the dimeric mercurous ion Hg_2^{+2} is the complex of Hg^{+2} with a mercury atom having equilibrated their electronic density to become symmetric. It is remembered that monomeric HgX is only known in the gaseous state (like the diatomic molecules MgX, BaX and ZnX studied by molecular spectroscopists). The solids containing linear tetratomic XHgHgX disproportionate readily (in particular, an excess of X^- produces HgX_4^{-2} and metallic mercury). A ligand somewhat comparable to the mercury atom is $SnCl_3^-$ known from the dark-red trigonal-bipyramidal $Pt(SnCl_3)_5^{-3}$ and yellow *cis*-$Pt(SnCl_3)_2Cl_2^{-2}$ containing the quadratic chromophore $Pt(II)Cl_2Sn_2$. Quite recently[138] AsF_6^- salts have been prepared under anhydrous conditions of the new catenated ions Hg_3^{+2} and Hg_4^{+2}. These cations are open chains and not like the square Te_4^{+2}.

D. Lead(II) and Other Gillespie-unstable Systems

The isoelectronic series Hg, Tl^+, Pb^{+2}, Bi^{+3}, . . . have the groundstate belonging to the electronic configuration terminating $6 s^2$. The first excited configuration $6s6p$ contains four levels $^3P_0$, $^3P_1$, $^3P_2$ and $^1P_1$. Allowed electric dipole transitions are possible to $^3P_1$ and $^1P_1$ and have considerable oscillator strengths P around 0.5 and 1.5. The wave-numbers increase regularly; $^3P_1$ is situated at $39\,412$ cm^{-1} in Hg (giving the sharp absorption line from a 10 cm cell containing vapour of a drop of mercury, which can be used for calibration of spectro-photometers in the ultra-violet) at $52\,393$ cm^{-1} in Tl^+ and $64\,391$ cm^{-1} in Pb^{+2}. In compounds[69] the transitions to two excited states remain roughly constant in the region 30 000 to 50 000 cm^{-1} for a given set of ligands of Tl(I), Pb(II) and Bi(III). Duffy and Ingram[139] actually use the variation of band positions to classify the ligands much in the same way as the nephelauxetic series. The post-transition group halides show reflection[140] and solution spectra dominated by these inter-shell transitions rather than by electron transfer.

Though the original spectral assignments assumed spherical symmetry, the stereochemistry of these central atoms seems unusually distorted. It is a great contrast that the various modifications of yellow PbO have the short Pb—O distances at one side only, as if the rear side contained a bulky lone-pair, but PbS, PbSe and PbTe are cubic (NaCl type) low-energy-gap semiconductors (large crystals show metallic luster) of an aspect very different from the colourless isotypic BaS, BaSe and BaTe. It cannot be excluded[67] that the lead atoms in PbS are statistically distributed in the eight directions along the trigonal axes or in the six directions along the tetragonal axes in a way keeping the six Pb—S distances highly different on an instantaneous picture.

Anyhow, the distorted stereochemistry of such post-transition group compounds is described by Gillespie[141] as lone-pairs occupying the positions of ligands, as was originally proposed by Sidgwick for explaining the resolution of sulphonium ions with three different substituents $RR'R''S^+$ in optically active enantiomers as if they were tetrahedral with one lone-pair and three ligands. We return to this problem below, but an alternative explanation was suggested by Orgel[142] that $6s$ and one of the empty $6p$ orbitals are mixed in the MO of the distorted chromophore. This is exactly an expression of the spontaneous loss of a centre of inversion.

Lead(II) chemistry is comparable to Hg(II) as far goes a (less pronounced) affinity to iodide, sulphide and sulphur-containing ligands, but not to amines, phosphines and cyanide. It can be argued that strongly σ-bonded ligands would make the $6s$ orbital anti-bonding. Organo-lead chemistry is concentrated on tetrahedral lead(IV) compounds such as the herostratically famous anti-knocking agent $Pb(C_2H_5)_4$, and in view of the explosive character of the yellow oil $PbCl_4$ (thermally much less stable than salts of $PbCl_6^{-2}$ with large univalent cations) it is surprising that halides and carbanion ligands co-exist in *colourless* $Pb(C_2H_5)_2Cl_2$ and $Pb(C_2H_5)_2Br_2$.

In aqueous solution, N is low and most frequently 3. The amphoteric dissolution of $Pb(OH)_2$ in excess of OH^- is now known to produce $Pb(OH)_3^-$, an interesting contrast to $Be(OH)_4^{-2}$ formed by a much smaller central atom. The moderately soluble $PbCl_2$ and PbI_2 are known to dissolve in an excess of Cl^- to $PbCl_3^-$ or of I^- to PbI_3^-. It is not certain whether $PbCl_3^-$ binds additional chloride ligands; the absorption spectrum of Pb(II) shifts marginally[143] as a function of the concentration of strong HCl.

Crystal structures containing the s^2 type cations are sometimes cubic of types such as NaCl, K_2PtCl_6 and $SrTiO_3$ with $N = 6$ and in other cases highly distorted. On the whole, oxide and fluoride tend to express the distortions. Thus, TeF_5^- is square-pyramidal like the isoelectronic IF_5 and XeF_5^+ and a text-book illustration of Gillespie-behaviour as an octahedron with a lone-pair and five ligands, whereas $TeCl_6^{-2}$ and $TeBr_6^{-2}$ exist (like gaseous XeF_6). Couch, Wilkins, Rossman and Gray[144] discuss the weak band separations of the excited levels $^3P_1$ and $^1P_1$ of TeX_6^{-2} corresponding to a weak orthorhombic distortion on an instantaneous picture. Pearson[145] prefers to speak about second-order Jahn-Teller effect rather than Gillespie-instability. However, the occurrence of second-order Jahn-Teller distortions is a numerical question of whether the vibronic coupling with excited Ψ of opposite parity is stronger than the restoring forces of the groundstate. This is a subtle question to be further discussed in the quantum-chemical section.

III. Electric Dipolar Polarizability and the Approximately Additive Molar Refractivities

Whereas the hard-hard interactions in Pearson's Dual Rule essentially are Coulombic, the soft-soft interactions frequently invite the comment that they represent *polarizability*. This word has connotations for the chemists which are not very different

from soft bases and anti-bases including species such as silver(I) and mercury(II). However, the situation is not as clear-cut for the physicist.

In the visible and ultra-violet region of frequencies ν (the wave-number σ is ν/c) the electric dipolar polarizability α is connected with a summation over all the k excited states to which the transition has the oscillator strength P_k:

$$\alpha = \sum_k \frac{P_k \text{ bohr}^3 \text{ hartree}^2}{h^2(\nu_k^2 - \nu^2)} \tag{27}$$

where 1 bohr = 0.529 Å and 1 bohr^3 = 0.1482 Å^3, and 1 hartree = 2 rydberg = 27.2 eV = 219 500 cm^{-1}. The oscillator strength of an absorption band is given Eq. (21). The best experimental technique for evaluating α is to measure the refractive index n of a pure compound for which the density d and molar weight M is known, *via* the *molar refractivity*

$$R = \frac{n^2 - 1}{n^2 + 2} \cdot \frac{M}{d} = \frac{4\pi}{3} N_0 \alpha \tag{28}$$

Since N_0 is Avogadro's number, 1 cm^3 of R is equivalent to 0.392 Å^3 of α. The factor $(n^2 - 1)/(n^2 + 2)$ multiplying the molar volume (M/d) is close to zero for gases and approaches 1 for highly refractive materials. It is even argued[146] according to a hypothesis by Herzfeld from 1927 that a material would become metallic at such a high pressure where the molar volume has been reduced to R from Eq. (28).

Organic chemists tend to describe R of a molecule as a sum of bond properties whereas inorganic chemists prefer a sum of characteristics of the atomic entities. Fajans[147] reviewed this subject and emphasized that one cannot describe ionic salts by an additive R or α of the individual ions. Salzmann and the writer[148] looked into this question, and though we admit that one cannot obtain a perfect additivity, we also believe that it is possible to choose a set of constituent polarizabilities much closer to additivity than maintained by Fajans.

Table 2 gives the α values for a series of gaseous, liquid and crystalline compounds. For convenience, the subtraction $-\nu^2$ in the denominator of Eq. (27) has been neglected for yellow light from a sodium lamp. Table 2 includes values for species in aqueous solution mainly derived from careful measurements by Heydweiller[149]. Though α for water is low, the molar weight M is also low, and (α/M) is 0.081, between the values 0.05 for iodide and 0.085 for chloride. However, besides this question of experimental uncertainty, a much more serious problem is how α can be distributed on the *individual ions*. This is a general problem in physical chemistry that we only have collective properties of neutral substances. In such a case, an additivity rule will be satisfied as well or as unsatisfactorily if all values for the ion with the charge z are added an arbitrary constant C multiplied by (the positive or negative) z. This ambiguity is usually circumvented by rather ingenious arguments. For instance, molar conductivities cannot conceivably be negative. Given this, one may try to look for an ion with very small conductivity, or one may apply Stokes' law for motion in a viscous fluid and assume a proportionality between reciprocal ionic radii and the ionic conductivity. It turns out that the "Stokes radius" of Na$^+$

Table 2. Molar refractivities in Å^3 derived from measurements of refractive indices

H (calc.)	0.67		$(HO)_2AsO_2^-$ (cryst.)	6.5
H($-$I)	1.8		SeO_4^{-2} (cryst.)	6.5
He (gas)	0.205		SeF_6 (gas)	5.25
Li^+ (gas)	0.027		Br($-$I) (cryst.)	4.2
Li(I)	0.03		(aq.)	4.6
Be(II)	0.0		(C$-$Br)	3.4
Be_2C (crystal)	3.21		Kr (gas)	2.46
C (diamond)	0.83		Rb(I)	1.9
CH_4 (gas)	2.60		Sr(II)	1.4
CO_2 (gas)	2.65		Y(III)	1.5
CO_3^{-2} (cryst.)	4.0		Zr(IV)	1.8
CO_3^{-2} (aq.)	4.4		Nb(V) (Nb_2Cl_{10})	6
NO_3^- (cryst.)	3.9		MoO_4^{-2} (aq.)	10
NO_3^- (aq.)	4.0		Rh(III)	2.5
NH_3 (gas)	2.26		Pd(II) $(PdCl_4^{-2})$	4
H_2O (aq.)	1.458		Ag(I)	2.3
O($-$II) (BeO)	1.25		Cd(II)	1.7
(SiO_2)	1.3		In(III)	1.6
(B_2O_3, Al_2O_3)	1.35		Sn(IV) $(SnCl_4)$	3.4
(MgO)	1.55		$(SnBr_4)$	5.9
(CaO)	1.95		$SbCl_5$ (liq.)	17.3
(Li_2O, SrO)	2.05		Sb(V)	4
F($-$I) (cryst.)	0.9		TeF_6 (gas)	5.78
(aq.)	0.7		I($-$I) (cryst.)	6.3
(CF_4)	0.7		(aq.)	7.1
Ne (gas)	0.39		(C$-$I)	5.4
Na(I) (cryst.)	0.3		Xe (gas)	4.0
(aq.)	0.4		Cs(I)	2.9
Mg(II)	0.2		Ba(II)	2.4
Al(III)	0.0		La(III)	2.2
PO_4^{-3} (cryst.)	6.0		Ce(III)	2.4
$(HO)_2PO_2^-$ (cryst.)	5.3		Ce(IV)	4
SO_4^{-2} (cryst.)	5.1		Pr(III)	2.3
(aq.)	5.3		Nd(III)	2.2
SF_6 (gas)	4.47		Sm(III)	2.2
Cl($-$I) (cryst.)	3.0		Gd(III)	1.9
(aq.)	3.2		Yb (III)	1.6
(CCl_4)	2.6		Hf(IV)	1.7
ClO_4^- (cryst.)	4.9		WO_4^{-2} (cryst.)	7.9
(aq.)	5.0		OsO_4 (gas)	6.3
Ar (gas)	1.63		PtF_6^{-2} (cryst.)	7.0
K(I)	1.2		$PtCl_6^{-2}$ (cryst.)	21
Ca(II)	0.9		$PtBr_6^{-2}$ (cryst.)	30
Sc(III)	1.1		Hg (gas)	5.0
TiO_2 (crystal)	4.65		Hg(II) (aq.)	2.9
$TiCl_4$ (liq.)	15.0		$(HgCl_2)$	3.4
Ti(IV)	2 to 4		Tl(I) (aq.)	4.3
V(III)	1.4		(cryst.)	4 to 6
Cr(III)	1.6		Pb(II) (aq.)	4.3
CrO_4^{-2} (aq.)	9.8		(cryst.)	3.7 to 5
Mn(II)	1.3		Bi(III) (aq.)	4
Fe(II)	1.1		(cryst.)	3
Fe(III)	2.2		Th(IV)	2.7
Co(II)	1.1		U(III) (UCl_3)	3.2
Ni(II)	1.0		U(IV)	3.2
Cu(II)	1.0		UO_2^{+2}	5.6
Zn(II)	0.9		U(VI)	3
$GeCl_4$ (liq.)	12.5		Pu(IV)	3.4
Ge(IV)	2			

is larger than of K^+ in aqueous solution, which can be explained away with an extended hydration shell of Li^+ and Na^+, but at least ions like $As(C_6H_5)_4^+$ and $B(C_6H_5)_4^-$ are sufficiently large. If one calculates molar volumes of dissolved salts, it is very difficult to escape the conclusion that Mg^{+2} in anhydrous $MgSO_4$ produces a negative contribution. This is not aberrant because the six ligands in $Mg(H_2O)_6^{+2}$ formed may very well have a lower molar volume each (say 10 cm^3) than liquid water known to have a rather uneconomic way of packing.

In our case it is clearly desirable not to have negative α values though Fajans analyzed Heydweiller's results in a way where Mg(II) in aqueous solution has $\alpha = -0.75$ Å^3 and Al(III) $\alpha = -1.2$ Å^3. It cannot be excluded *a priori* that such negative values might be connected with a decrease 0.2 Å^3 of α of each of the six ligated molecules, or in a more collective alternative, that *all* the water in a 1 M solution of these cations has α decreased by 0.02 Å^3. However, our analysis[148] shows that it is more reasonable to add Cz to the values derived by Fajans with $C = 0.4$ Å^3. A comparison with refractive indices of crystalline compounds allows zero-points to be fixed by assuming vanishing α of Be(II) and Al(III). For practical purposes, since the additivity anyhow does not work better than within one or two-tenths Å^3 one may also neglect the contribution of Li(I).

In a few cases, the wave-function Ψ of a monatomic entity can be used for calculating α, *e.g.* 4.5 bohr^3 for the hydrogen atom, or 0.205 Å^3 for the helium atom in agreement with the experimental value. Gaseous H$^-$ does not have a Hartree-Fock function stable relative to spontaneous loss of an electron, and it is necessary to introduce correlation effects in order to calculate α which is said to be 31 Å^3. The value 1.8 Å^3 for H($-$I) in Table 2 derives from NaCl-type LiH, NaH and KH. The anion $B_2H_6^{-2}$ has $\alpha = 6.3$ Å^3 to be compared with the isoelectronic C_2H_6 4.47 Å^3. Since CH_4 has $\alpha = 2.6$ Å^3, each hydrogen can at most contribute 0.65 Å^3 in what case the C–C bond in ethane contributes 0.57 Å^3. This would be slightly more than 0.42 Å^3, half the value for diamond. Much the same problem as for hydrides occur for *oxides*. Gaseous O^{-2} looses spontaneously an electron, and the oxides only occur because of a stabilizing Madelung potential. There is no doubt that orange Cs_2O would show a large α, and this compound is readily oxidized to the peroxide Cs_2O_2 or the superoxide CsO_2 corresponding to the low ionization energy of oxide in the weak Madelung potential prevailing in Cs_2O. It can be seen from Table 2 that the lower limit of α for an oxygen atom in a compound is some 1.25 Å^3 as also found by comparison with NO_3^-, SO_4^{-2} and ClO_4^- with vanishing contributions from the central atom.

The main conclusion from Table 2 is that the dramatic decrease of α with increasing ionic charge z for a given isoelectronic series *does not always occur in compounds*. It is true that α decreases monotonically in the series O($-$II), F($-$I), Ne, Na(I), Mg(II) and Al(III) but far less at the beginning than between the infinite value for O^{-2} and a value for gaseous F$^-$ not far from 2 Å^3. However, there is a shallow minimum close to the bivalent ion in the series Br($-$I), Kr, Rb(I), Sr(II), Y(III) and Zr(IV) and close to the trivalent ion in the series I($-$I), Xe, Cs(I), Ba(II), La(III) and Ce(IV). The important point is not the exact position of this flat minimum, but the completely different behaviour from the gaseous ions. One might have suspected, once more, that α is a decreasing function of increasing fractional atomic charge[4, 36, 39] of the atom. However this cannot be the whole truth because α is connected with the positions and oscillator strengths of excitations according to Eq. (27). One reason for the turning around, α increasing with the oxidation state, *e.g.* from Ca(II) to Cr(VI), is the electron transfer bands[45]. When they have sufficiently low wave-numbers, it is no longer possible to make a significant separation between contributions to α from the central atom and from the ligands, and fluoro complexes have distinctly lower α than chloro complexes. Thus, the two known electron transfer bands of $PtCl_6^{-2}$ provide $\alpha = 6.2\ \text{Å}^3$ and the four transitions in $PtBr_6^{-2}$ 10.8 Å^3[148]. These large contributions must in part spend some of the halide polarizability.

With exception of the hydrogen atom and the $6s^2$ isoelectronic series (the mercury atom, Tl(I), Pb(II) and Bi(III) compounds) most systems have the majority of the effective (P_k/ν_k^2) above the first ionization energy, in the *continuum*. Thus, the noble gases would have the observed α values if all the P_k was concentrated at $\nu_k =$ the ionization energy with the numerical values of $P_k = 0.84$ for helium, 0.99 neon, 2.00 argon, 2.24 krypton and 2.59 for xenon. Since helium contains two $1s$ electrons, and the four other noble gases have six p electrons in their loosest bound shell, these figures indicate that most of the α derives from transitions in the continuum.

Among the unexpected results compiled in Table 2 is that α for M(II) of the $3d$ group decreases smoothly from Mn(II) to Zn(II) and that α for M(III) of the $4f$ group also decreases slowly from Ce(III) to Yb(III). It is possible[148] to discuss the contributions of the individual shells to α with the negative result that each of the $3d$ electrons of Zn(II) or each of the $4f$ electrons of Yb(III) and Hf(IV) contribute less than 0.1 Å^3 to α. The theoretical estimate of the $5p$ contribution is 1 Å^3 in the two latter ions.

For the chemist, there is no doubt whatsoever that the softness of the central atoms do not follow the α values, though this is valid for the halides. Thus, Cs(I) is more polarizable than Ag(I), Ba(II) more than Cd(II) and La(III) more than In(III) in disagreement with Pearson's classification. Said in other words, the chemical bonding involves far stronger perturbations than the linear electric fields inducing the α-polarizability.

IV. Quantum-chemical Comments

Quantum mechanics is approximately 50 years old, and it is worthwhile to ask what it has done for chemistry[150]. Questions such as Pearson's Dual Rule are among the most difficult to treat in quantum chemistry with an even remote hope of significant results. First of all, the energy differences of interest to complex chemists are exceedingly small compared with the energy of Z electrons being bound to the nucleus with charge $+Ze$ having the order of magnitude $-Z^{2.4}$ rydberg. For this purpose, it is useful to compare various units of energy:

$$1000\ \text{cm}^{-1} = 0.1239\ \text{eV} = 2.85\ \text{kcal/mole} = 11\,920\ \text{joule/mole} \qquad (29)$$
$$\equiv 2.09\ \text{powers of ten in equilibria constants at 25 °C}$$

Secondly, most quantum-chemical discussion assumes *fixed* internuclear distances producing what spectroscopists call *vertical transitions* and ionization processes studied in recent photo-electron spectrometry [151, 152] by analogy called *vertical ionizations*. The *adiabatic processes* allowing the internuclear distances to vary represent a tremendous complication in a polyatomic system containing Q nuclei because the *potential hypersurfaces* are the eigenvalues of the total electronic wavefunctions Ψ as a function of the $(3 Q - 6)$ Cartesian coordinates remaining as degrees of freedom after separation of three translational and three rotational variables. Because of paper having two dimensions, most people are accustomed to the potential curve of a diatomic molecule with *one* internuclear distance. The activation energy of a reaction is the maximum height of the easiest walk on the potential hypersurface between one minimum representing the groundstate of the original reactants and another minimum representing the products. The intermediates are relative minima and transition states (usually) saddle points on the hypersurface.

The *preponderant electron configurations* classify correctly the (up to 400 levels) compiled by Charlotte Moore in the National Bureau of Standards Circular "Atomic Energy Levels". Though this does not mean that a certain amount of configuration interaction does not take place in the Ψ of each of these levels of monatomic entities, it is possible to extend this classification to the approximate spherical symmetry of compounds containing a partly filled $4f$ shell. Actually, it is even possible to define the *oxidation state* [4] for p and d group compounds in the cases where the distribution of excited levels observed in the absorption spectrum is sufficiently characteristic to indicate unambiguously the preponderant configuration of the groundstate. In general, certain oxidation states such as $3d^3$ Cr(III) or $3d^8$ Ni(II) are readily recognized, but one can run into problems with strongly coloured compounds of ligands capable of adding electrons (such as dipyridyl or quinones) or loosing electrons, as certain conjugated sulphur-containing ligands. In such cases, the ligands are said to be collectively reduced (as is also true for the excited levels of electron transfer to certain conjugated ligands, increasing the oxidation state of the central atom by one unit) or collectively oxidized (as is also true for the usual electron transfer bands, where the oxidation state of the central atom is decreased one unit in the excited level). A few simple ligands (NO^+, NO, NO^- or O_2, O_2^-, O_2^{-2}) present comparable problems. It is clear that the attribution of oxidation states to closed-shell colourless systems is more arbitrary [4] and done by extrapolation and analogy. It helps that most ligands are *innocent* having closed shells for the purpose of finding the preponderant configuration of the central atom, as has been learned from the transition groups. Ammonia is a neutral ligand isoelectronic with CH_3^- and with BH_3^{-2} which seems to occur [4] in $Mn(CO)_5BH_3^-$ becoming an octahedral $3d^6$ Mn(I) complex in spite of the fact that BH_3 is a well-known anti-base. Seem from the point of view of oxidation state, there is never any possibility of *neutral* CH_3 or H as ligand (this does not mean the same thing as "functional group" to the organic chemist) but the choice may be between CH_3^- and CH_3^+ or between H^- and H^+ in cases such as $P(CH_3)_3$, PH_3, $S(CH_3)_3^+$, $Te(CH_3)_3^+$ and TeH_2 which may contain P(–III) or P(III), S(–II) or S(IV) and Te(–II) or Te(IV) and Te(II). One of the weak points of this argumentation is that $N(CH_3)_4^+$ and $P(CH_3)_4^+$ would have the choice between N(V) and P(V) with carbanion ligands or N(–III) and P(–III) with carbonium ligands. The horrible truth may be that these central atoms do not possess oxidation states. It is preferable[4] to describe CH_4 as a carbon(IV) hydride (in view of the substitution by halides in $CH_{4-n}X_n$) but NH_3 as N(−III) and H(I) in view of the ready protonation. Besides the spectroscopic properties of diamagnetic ($S = 0$) compounds, there is a second line of argumentation which *sometimes* may help defining the oxidation state, viz. the preferred stereochemistry[153]. For instance, when a platinum atom is surrounded by one hydrogen, one chlorine and two phosphorus (from phosphines) in a *plane* it is strong evidence for $5d^8$ Pt(II) whereas $5d^{10}$ Pt(0) would be expected to be tetrahedral. The chemist would expect the question of adapting CH_3^- and H^- with an electron pair more than CH_3^+ and H^+ to be decided by the relative electronegativity of the central atom considered and carbon (or hydrogen). Anyhow, all available spectroscopic and stereochemical evidence points to d group complexes choosing H^- and CH_3^-[4]. There is one situation where it is certain that the electronegativities are identical, the *catenation* in symmetrical species such as Hg_2^{+2}, H_3CCH_3, $(OC)_5MnMn(CO)_5$, $O_3SSO_3^{-2}$, $O_3SSSSO_3^{-2}$ or F_5SSF_5. It is not particularly informative to think about ethane heterolytically dissociated to CH_3^- and CH_3^+ or the dimeric manganese carbonyl to $Mn(CO)_5^-$ and $Mn(CO)_5^+$ and it is tempting to use a description related to Fajans' quanticules, to reserve a pair of electrons to the homo-atomic bond, and to let ethane be two CH_3^+ connected with such a pair, etc. This quanticule formulation can be extended to other bonds between two metallic elements of comparable electronegativity, such as $(OC)_5ReMn(CO)_5$, $(OC)_4CoMn(CO)_5$ and $ClSn(Co(CO)_4)_3$. This approach can go very far; the ligand $SnCl_3^-$ isoelectronic with $SbCl_3$ and $TeCl_3^+$ can be suspected for catenation between Sn(IV) and Co(I) in $Cl_3SnCo(CO)_4$ comparatible with the stereochemistry of tetrahedral Sn(IV) and trigonal-bipyramidal $3d^8$ Co(I), and it would not be perfectly ludicrous to regard the robust cobalt(III) complexes $Co(NH_3)_5NO_2^{+2}$, $Co(NO_2)_6^{-3}$ and $Co(NH_3)_5SO_3^+$ as catenation by six electron pairs of dioxo and tri-hydrido complexes of nitrogen(V) and sulphur(VI) trioxide. However, this description early meets its limits; it is not generally true that each chemical bond employs exactly two electrons. Seen from the point of view of functional groups, a chlorine atom bridging two other atoms is Cl^+ like N^+ in betaïne, and the central iodine in IAg_3^{+2} would be I^{+2}. It may be remembered[154] that PAg_6^{+3} and $TeAg_8^{+6}$ also exist. If catenation occurred in the examples, chlorine(III) and iodine(V) would rather astonishingly appear. It should be emphasized that we are not discussing the "resonance between valence-bond-structures" which has problems enough[69] but the optimized selection of oxidation states for the classification of the preponderant configuration of the groundstate. It may not always be possible to make such a selection.

The MO (molecular orbitals) of a polyatomic system are one-electron wave-function Ψ which can be used as a (more or less successful) result for constructing the many-electron Ψ as an anti-symmetrized Slater determinant. However, at the same time the Ψ (usually) forms a preponderant configuration, and it is an important fact[67] that the *relevant symmetry* for the MO may not always be the point-group determined by the equilibrium nuclear positions but may be a higher symmetry. For many years, it was felt that the mathematical result (that a closed-shell Slater determinant contains Ψ which can be arranged in fairly arbitrary new linear combinations by a unitary transformation without modifying Ψ) removed the individual subsistence of the MO. This argument neglects the fundamental difficulty that Ψ is *not* exactly a Slater determinant and that the main purpose of MO is to construct preponderant configurations which one can excite by changing the MO of an electron (or in ground-states with $S \geq 1$ by changing to other levels belonging to the ground configuration, as known from O_2 and many d group complexes) and *ionize* by removing an electron from one of the MO. It has never been necessary to convince atomic and X-ray spectroscopists that no orbital in the krypton atom is a mixture of $1s$ and $4p$ as one would argue from the unitary transformation. However, the *penultimate* MO with higher ionization energy I than the loosest bound MO were not taken too seriously before the photo-electron spectra of gaseous molecules [151, 152] provided experimental results to compare with the calculated I values.

In isolated molecules, it is usually assumed that the MO are satisfactorily approximated by LCAO (linear combination of atomic orbitals). This hypothesis is supported to a great extent by careful calculations on diatomic and other small molecules, but it cannot be proved generally. Though the chemical bonding energies are rather negligible compared with the total binding of electrons in each atom (excepting hydrogen) the outermost electrons are "soft" in a certain sense, being flexible by perturbations from adjacent atoms. First of all, it is beyond doubt that the radial functions of each AO contract under the influence of increasing fractional atomic charge [4, 39, 67] and secondly, the kinetic energy operator in the Schrödinger equation produces a contraction of the AO constituents of bonding MO and an expansion of anti-bonding MO with a nodal surface between the two atoms, as first pointed out by Ruedenberg [155]. It has been suggested [67] that the "ligand field" effects are mainly determined by the kinetic energy operator in the bond region.

It is only in crystals that one frequently attempts to solve the Schrödinger equation directly by finding the one-electron orbitals as eigen-functions of an effective potential $U(x, y, z)$ for instance in the "augmented plane-wave method" where the Ψ are kept orthogonal on the inner shells of each atomic core but correspond to free electrons outside the cores where U is constant. Recently, Slater and Johnson [156, 157] have proposed the "multiple-scattering Xα method" for oligo-atomic entities with U chosen according to certain rules. The agreement [158] with I values in photo-electron spectra and with excited levels is better than with many more sophisticated approximations. It is clear that the LCAO hypothesis plays a minor role in such direct solutions of the Schrödinger equation.

In molecules, *mixing of l values of the same atom* can occur in MO of a definite symmetry type if permitted by group theory applied to the point-group of the relevant symmetry. This mixing might be called "hybridization" but the writer discourages this word which has a different meaning [48, 69] in valence-bond theory. In linear molecules without centre of inversion (such as CO, HF, NNO, NCS⁻ and HCCCl) the $2s$ orbitals of a given carbon, nitrogen, oxygen or fluorine atom can be mixed with the cylindrically symmetric $2p\sigma$ orbitals of the same atom, but not with the $2p\pi$ orbitals having a node-plane containing the linear axis. In linear molecules *with* centre of inversion (such as N_2, OCO, NCN⁻², NNN⁻, OUO⁺², HCCH and NCHgCN) the σ orbitals have to choose between even and odd parity and are called σ_g and σ_u (g = gerade, u = ungerade in German). *If* the atom considered has its nucleus at the centre of inversion (this is the case for the central atom in the case of an odd number of nuclei) only l values of the same parity can be mixed. Thus, the s orbital can mix with the cylindrically symmetric $d\sigma$ orbital having the angular function proportional to $(3z^2 - r^2)$ in linear $HgCl_2$ and $Hg(CN)_2$ and also in complexes such as $Ag(NH_3)_2^+$, $Hg(NH_3)_2^{+2}$ and $Hg(CH_3)_2$ which can be considered, to a good approximation, to contain the linear chromophores $Ag(I)N_2$, $Hg(II)N_2$ and $Hg(II)C_2$. The atoms having their nuclei situated on general points of the linear axis (outside the center of inversion) can mix odd and even l values in the MO of σ_g or σ_u symmetry. Thus, the orbital most easy to ionize in N_2 of symmetry σ_g has mainly $2p$ and some $2s$ character. The squared amplitudes (which add to exactly 1 because of the absence of intra-atomic overlap integrals) seem to be 10 to 25 percent *vs.* 90 to 75 percent in most cases of $2s$ and $2p$ mixing. The reason why the coefficients to LCAO are so moderate is that the diagonal elements of $2s$ and $2p$ orbitals at the same atom differ 10 to 15 eV, as also confirmed by photo-electron spectra[151].

Another typical case of l mixing conserving the parity is the mixing[133] of the totally symmetric (a_{1g}) s and $d(3z^2 - r^2)$ orbitals in the point-group D_{4h} characterizing quadratic chromophores with the four ligating atoms in the xy-plane. However, for our purposes, the mixing (even to a small extent) of odd and even orbitals on the same atom is far more important. The point is that both odd and even ψ have even parity of their square ψ^2 representing the electronic density. A mixture of odd and even ψ does not have a well-defined parity and can concentrate on one side of a plane containing the nucleus of the atom considered. Some tetra-atomic molecules containing a central atom and three identical ligating atoms adapt the highest possible symmetry, D_{3h} with the ligands in an equilateral triangle (such as BF_3 or NO_3^-) but other become pyramidal (C_{3v}) like NH_3, NF_3, SO_3^{-2}, ClO_3^-, IO_3^- and XeO_3. However, these species have *holohedrized symmetry* D_{3d} like a trigonal anti-prism, whereas AsH_3 has almost right angles HAsH. If they were right, the point-group would still be C_{3v} but the holohedrized symmetry[67, 82] is O_h like the chromophore $Pt(IV)C_3O_3$ in *fac*-$Pt(CH_3)_3(H_2O)_3^+$. By the same token, TeH_2 with a right angle has the same point-group C_{2v} as H_2O, but the holohedrized symmetry is D_{4h}. There is no particular reason to believe[67, 68] that the bond angles in pyramidal molecules are determined by the participation of p orbitals in $s-p$ hybridizations. Nevertheless, it is not easy to explain why some polyatomic systems are much more apt than others to deviate from the highest available symmetry. Exactly like the first-order Jahn-Teller effect, the Gillespie-distortions may either be *static* (seen on crystal structures) or *dynamic* (to be observed on instantaneous pictures obtained by spectroscopic techniques). Many post-transition group complexes are highly suspect for dynamic Gillespie-instability (XeF_6 is a famous case, though we must add that xenon(VI) is moderately hard, in view of the existence of XeO_3 and XeO_3X^- with X = OH, F and Cl) and we give the arguments above why copper(II) and palladium(II) are Gillespie-unstable though the deviations go along the z-axis in C_{4v} of $Cu(NH_3)_5^{+2}$ in the former case and along the x- and y-axes in C_{2v} of $Pd(II)N_2O_2$. Gillespie[141] does not believe that the d-electrons in general behave like lone-pairs in the transition group compounds, but he adds that a beginning tendency toward distortion can be seen at the end of the d groups. It is perhaps significant that the kinetically more rapid central atoms Au(III), Cu(II) and Pd(II) at the same time are more oxidizing and more readily deviate from quadratic chromophores. However, it must be admitted that the vibrational energy levels (obtained by Raman and infra-red spectroscopy) of quadratic species such as $PdCl_4^{-2}$ and $PtBr_4^{-2}$ do not show broadening or separations of frequencies degenerate in the point-group D_{4h}. This criterium would not apply to almost octahedral TeX_6^{-2} and XeF_6. Still, the band intensities of internal $3d^9$ transitions in Cu(II) and $4d^8$ in Pd(II) given in Table 1 are far too large to convince about an unconditional fidelity to the point-group D_{4h}. Again, this may be a question of time-scale[67].

Seen from a conventional LCAO point of view, the condition for the existence of complexes of CO, olefins such as C_2H_4, and of NO^+ is *synergistic back-bonding*. It seems well established from force constants obtained by vibrational spectra[159] that back-bonding occurs in the isoelectronic ($3d^{10}$) series $Ni(CO)_4$, $Co(CO)_4^-$ and $Fe(CO)_4^{-2}$ and becomes more important in direction of the negative oxidation numbers. In a way, it is curious that CO having one of the highest $I = 14.1$ eV known for a diatomic molecule and the first excited levels above 6 eV should be a simultaneous lone-pair σ-donor and π^*acceptor. Until the preparation of complexes such as $Ru(NH_3)_5N_2^{+2}$ it was a standing question why N_2 would not work equally well. Though iridium(III) forms $Ir(CO)Br_5^{-2}$ the fact of CO complexes being concentrated between M(–II) and M(II) (a striking case is the orange $5d^6$ $Ta(CO)_6^-$ being the only known monomeric tantalum complex with an oxidation number below 4) is taken as an indication of back-bonding, and quantum-mechanical calculations provide sufficiently large non-diagonal elements of the effective one-electron operator[160].

The situation with *phosphines* is much more difficult to understand. Kruck[55] demonstrated that PF_3 is the only ligand more effective than CO to stabilize negative oxidation states of d group elements (excepting NO^+ if one argues that yellow $Co(NO)(CO)_3$, red $Fe(NO)_2(CO)_2$ and green $Mn(NO)_3(CO)$ all three are $3d^{10}$ systems). Colourless $4d^{10}$ $Ru(PF_3)_4^{-2}$, $Rh(PF_3)_4^-$ and $Pd(PF_3)_4$, $4d^8$ $Ru(PF_3)_5$, $5d^{10}$ $Os(PF_3)_4^{-2}$ and $Ir(PF_3)_4^-$ and mixed hydrido-complexes such as $4d^8$ $Rh(PF_3)_4H$ and $5d^8$ $Ir(PF_3)_4H$ are extreme cases, and not subject to the doubt[4] about reduced ligands as the strongly coloured dipyridyl complexes $M dip_3^{-z}$. Wilkinson previously prepared $Ni(PCl_3)_4$ and Bigorgne[161] after extended studies of vibrational spectra of $Ni(PR_3)_n(CO)_{4-n}$ and $Cr(PR_3)_n(CO)_{6-n}$ succeeded in preparing $Ni(PH_3)_4$. Secondary phosphines PR_2H tend to deprotonate to PR_2^- and this ligand readily forms bridges between two atoms of metallic elements[162]. Tertiary phosphines PR_3 form complexes of Co(II) and Ni(II) which can be prepared in non-aqueous solvents, and of typical soft central atoms in positive oxidation states such as Cu(I), Pd(II), Ag(I), Pt(II), Au(I) and Hg(II). However, the bidentate diphos = $(C_6H_5)_2PCH_2CH_2P(C_6H_5)_2$ stabilizes somewhat unusual oxidation states such as $5d^8$ iridium(I) Ir $diphos_2^+$. Another diphosphine-*ortho*-$C_6H_4(P(CH_3)_2)_2$ has recently been shown by Warren and Bennett[44] to stabilize high oxidation states such as Fe(IV), Ni(III), Ni(IV) and Cu(III). The corresponding diarsine was previously applied by Mann and Nyholm[40] with comparable results. Phosphite esters $P(OR)_3$ such as $P(OCH_3)_3$ and the sterically constrained $P(OCH_2)_3CCH_3$ were used by Verkade and Piper[163] to make diamagnetic chromophores such as trigonal-bipyramidal $Co(I)P_5$ and $Ni(II)P_5$ and octahedral $Co(III)P_6$. It is striking that these complexes are straw-yellow or colourless with the first excited levels in the ultra-violet showing a much later position in the spectrochemical series with larger sub-shell energy differences than the $3d$ group complexes of PR_3 characterized by their bright blue, violet and red colours.

One may, of course, argue that synergistic back-bonding occurs to the empty $3d$ orbitals of PF_3, PR_3 and $P(OR)_3$. However, this proposal is more difficult to defend than in the case of CO. The empty orbitals must have high energy in the inorganic and aliphatic ligands showing no excited levels before some 6 eV, and it is difficult to provide chemical bonding with large non-diagonal elements of the effective one-electron operator because of the longer M—P than M—C distance. The whole idea of $3d$ participation in phosphorus and sulphur originated in the belief that $N = 5$ in PF_5 or $N = 6$ in PF_6^-, PCl_6^- or SF_6 needs N orbitals for the chemical bonding and hence one or two orbitals (called "transargonic" orbitals by Pauling) besides $3s$ and $3p$. It is debatable whether spectroscopic evidence[164] exists for such $3d$ effects. An extrapolation of the electron transfer spectra backward from purple MnO_4^-, yellow CrO_4^{-2}, colourless VO_4^{-3}, ... or starting with $TiCl_4$ shows that already around calcium, the excitation energies are in excess of 9 eV and actually, it is rather molecules with smaller N and low I such as PR_3 and SR_2 which might conceivably have accessible empty $3d$ orbitals. Colton[165] suggests that the characteristic "soft" behaviour of iodide and bromide in noble metal complexes is due to back-bonding to empty $5d$ and $4d$ orbitals, respectively. Seen from the point of view of ultra-violet spectra[166, 167] this proposal is even less attractive, since the first excited levels are due to the configuration terminating $4p^5 5s$ in bromides and in krypton and $5p^5 6s$ in ionic iodides (5.5 eV) and xenon (8.5 eV) followed by $5p^5 5d$ levels at higher wave-numbers. The chemical argument would be that fluorides having no low-lying $3d$ orbitals show different properties. If stretched a bit, it might also illustrate why the dissociation energy of F_2 is lower than of Cl_2.

It is necessary to analyze what one means exactly with "$3d$ orbitals". If they have contracted radial functions, the overlap integrals with occupied ligand orbitals may be rather large. It is clear that one needs the two angular nodes characterizing $l = 2$ but that the principal quantum number 3 has no particular significance. From this point of view, the d orbitals are *polarization orbitals* also needed[168] for explaining the barrier toward inversion of NH_3. This molecule would be planar if the bonding involved s and p orbitals only. By the same token, the very low barrier toward free rotation in C_2H_6 can only be explained by f polarization orbitals with three angular nodes. The name "polarization orbitals" derives from the fact that an electric field $V = Cz$ changes the wavefunction Ψ_0 to $(\Psi_0 + CA\Psi_1)$ with appropriate units, where Ψ_1 is a linear combination of excited states with opposite parity, and the electric dipole moment of $CA\Psi_0\Psi_1$ represents the α values given in Table 2. Since the electric fields available in the laboratory, some 10^4 V/cm, are much smaller than the atomic unit of electric fields 530 MV/cm this first-order perturbation is a sufficient description. Hence, only the cross-term $\Psi_0\Psi_1$ enters the discussion.

It is by no means easy to say whether d polarization orbitals are needed in the quantum-chemical description of phosphorus and sulphur compounds. Because of the variational principle, one has to be exceptionally unlucky not to ameliorate an approximate Ψ when introducing a new free parameter. But the question is whether the d polarization orbitals are an essential aspect of the unknown (and somewhat Platonic) true Ψ. This is a very profound question related to the problem whether the *natural spin-orbitals* (introduced by Löwdin) having occupation numbers closely below 1 are those which define the preponderant configuration. It is now known that the natural spin-orbitals do not *need* to be symmetry-adapted but also that the convergence is *almost* as rapid if the constraint of (relevant?) symmetry-adaptation is introduced. The problem with natural spin-orbitals is that each set are a function of the approximate Ψ considered, but that function seems to converge so rapidly with increasing quality of Ψ that one is readily making the conceptual jump to talk about "the natural spin-orbitals of the true Ψ" as the limiting quantities.

It should be meaningful to ask whether the soft-soft interactions can be explained by the deformations of the MO alone. A specific point to discuss is whether the phosphines make σ bonds like amines, but just better with certain central atoms. This might be rationalized by the smaller bond angles CPC or FPF on the rear side of the phosphine, and by the bulky character of the lone-pair of the readily ionized (*i.e.* "vertically" reducing) phosphine corresponding to an orbital with low kinetic energy in a large confined volume. If one went very far, one might even catenate $X_3P(V)$ with *one* lone-pair to the central atom, arguing that the central atoms preferring phosphines favour catenation. This idea of an almost pure, single σ-bond would rather much be an anti-climax after so much talk about back-bonding. The σ-bond may involve more than an usual amount of deformation. In many ways, the question whether the M–P bond is performed by an essentially phosphorus $3p$ orbital or rather by a more general, cylindrically symmetric symmetric orbital is rather similar to the question whether the Mn–Mn bond in $(OC)_5MnMn(CO)_5$ is performed by a bonding linear combination of the $3d(3z^2 - r^2)$ orbitals of each atom, making the complex more or less $3d^7$, or whether the Mn–Mn bond is cylindrically symmetric without much relation to $3d$ orbitals. We remember that LCAO is no obligation derived from the Schrödinger equation.

On the other hand, genuine *more-electron effects* are known in chemistry. Ammeter and Schlosnagle[169] studied aluminium and gallium atoms incorporated in solidified noble gases. These two particular atoms have exceptionally strong *Van der Waals interactions* with the twelve adjacent noble-gas atoms because the un-paired $3p$ electron of Al does not have a much larger average radius than the occupied $3s$ orbital. Consequently, the neighbour atoms arrive closer than in a case like the $3s$ electron of a sodium atom, and the order of magnitude of chemical stabilization is 0.2 eV or 4 kcal/mole. The electron spin resonance spectrum of the aluminium or gallium atom show interesting consequences of the Jahn-Teller instability of a single p electron. Many other interactions are due to the interatomic correlation effects producing an energy minimum at long distance, such as the attraction between noble gas atoms explaining their condensation at low temperatures. It is curious to note that if the London dispersion forces between two iodine atoms (assuming an electric dipolar polarizability α close to 10 Å^3) are (rather dubiously) extrapolated to the equilibrium internuclear distance of I_2 the dissociation energy 1.5 eV of this molecule is obtained. Said in other words, it is difficult to make a sharp distinction between weak single bonds and the Van der Waals interactions between atoms containing partly filled p shells.

It is generally argued that the higher boiling points with increasing Z of the halide ($CH_4 - 164$, $CF_4 - 128$, $CCl_4 + 77$, $CBr_4 + 190 \,°C$) are due to increasing Van der Waals attractions. There is a general dependence of the boiling point on the molecular weight M of molecules (not associating or forming hydrogen bonds) but the fluorides generally fall low on this scale though they have the most pronounced internal separations of fractional atomic charges. When the octahedral molecules WF_6 form a liquid boiling at 17 °C but crystalline WCl_6 melts at 275 °C and boils at 347 °C, there is little doubt that one sees an effect of strong Van der Waals attractions in the latter case. One might argue that certain aspects of soft-soft interactions, and in particular the inorganic symbiosis, may suggest strong Van der Waals contributions, as when S_8 and HgI_2 are soluble in CS_2 but very little in most other solvents. It is worthwhile once more to mention that not only I^-, RS^- and R_3P are soft ligands, but also H^-, CH_3^-, CN^- and CO with far weaker capability of Van der Waals stabilization.

However, one of the rare distinct observable quantities related to Pearson's scale of softness. the shift of the I of the *inner shell* $2p$ of *potassium* by measurements[170] of photo-electron spectra of solid K^+ salts to be discussed below, and values given in Table 4, is a *relaxation effect*[171] due to the re-arrangement of the electronic density of the adjacent atoms concomitant with the ionization of the $K2p$ electron providing an ephemeral form of potassium(II) existing for some 10^{-15} sec Such chemical shifts amounting to some 2 eV cannot be incorporated in a one-electron operator containing the combined Hartree and Madelung potential, and are, technically speaking, correlation effects.

The writer has frequently expressed the view that induction from experimental data is necessary in quantum chemistry, and that some of our most unexpected and fascinating surprises come from new experimental techniques, such as the visible and ultra-violet spectra of transition group complexes 1950—65 or the photo-electron spectrometry starting around 1962. The present section is mainly intended to draw attention to the fundamental problems one may hope to solve in the future, and ambivalently appealing and frustrating queries: why is softness of anti-bases so frequently connected with mild deviations of the highest symmetry available (as exemplified by Cu(II), Pd(II) and Hg(II) and numerous cases of *trans*-influence) and why is the category H^-, CH_3^-, CN^-, CO, C_2H_4, C_6H_6, R_3P, RS^- and I^- *all* soft *and* symbiotic? Could it be catenation with *trans*-influence corresponding to the softest ligand attracting the electronic density from the other side of the central atom? It is realized that the heat of formation corresponding to soft-soft interactions normally is below 1 eV and $\log K$ is changed less than 16 units, as seen in Eq. (29). It would be very difficult for quantum chemistry to be *accurate* to this extent (in particular because the internuclear distances vary) and one has to be lucky to be *precise* in comparative studies. The last section reviews certain parametrizations directly related to Pearson's Dual Rule,

V. Softness Parameters, Hydration and Ionization Energies

A. The Dissolution of Gaseous Ions in Water

Classical thermodynamicists refrained from transporting electrically charged species across borders between two phases, and specifically from vacuo to an aqueous solution. The origin of this ascetic attitude is clearly the almost incredibly exact electroneutrality of macroscopic objects. One corollary of this restriction is that ions do not possess individual activities, only manifolds of ions having zero total charge (*e.g.* $2 K^+ + SO_4^{-2}$) however much Debye and Hückel use ion activities in their theory of long-distance interactions in dielectric solvents. Niels Bjerrum[172] wrote a paper about this paradox which extends very far; a well-known concept such as pH is not defined, strictly speaking, without reference to the surrounding salt medium and in particular to the ambient anions.

For the spectroscopist, there is nothing alarming about the ionization energy needed for removing one or more electrons form an atom (this process would be described by a classical physico-chemist as ΔH for forming a neutral gaseous mixture of ions M^{+z} and electrons at low pressure) even if the process is accompanied by the ion M^{+z} being transferred from vacuo to a dilute aqueous solution, *i.e.* we discuss the sum of positive ionization energies $I_1 + I_2 + I_3 + \ldots + I_z$ known from Charlotte Moore's tables (up-to-date version, see[173]) and the negative *hydration energy* for the transfer $M_{gas}^{+z} \rightarrow M_{aq}^{+z}$. We call this sum the *chemical ionization energy* I_{chem} for the process

$$M_{gas} = M_{aq}^{+z} + z \, e_{gas}^-. \tag{30}$$

Since the classical value for the standard oxidation potential E^0 of the normal hydrogen electrode is fixed at zero, the hydration energy of the proton has attracted special attention. Already in 1941, Jannik Bjerrum[6] gives the value -249 kcal/mole for this quantity which has been reviewed by Rosseinsky[174], revising the value to -260.5 kcal/mole. Knowing the dissociation energy 4.52 eV of H_2, the corresponding I_{chem} involved in the hydrogen electrode is 4.5 eV. Hence, it becomes attractive to evaluate I_{chem} for oxidation processes (not involving gases, where the ΔS has to be taken into account, nor solids with constant activity such as metallic elements) of the type $Fe(H_2O)_6^{+2} \rightarrow Fe(H_2O)_6^{+3} + e^-$ ($E^0 = +0.75$ V) or $Eu_{aq}^{+2} \rightarrow Eu_{aq}^{+3} + e^-$ ($E^0 = -0.4$ V) as the expression

$$I_{chem} = E^0 + 4.5 \text{ eV} \tag{31}$$

being 5.25 and 4.1 eV, respectively. It is conceivable that the difference of entropy ΔS cannot be consistently evaluated for processes of the type Eq. (30) and this is one reason why it is not perfectly clear whether the 4.5 eV represents ΔH or ΔG, and also why it might be situated somewhere between 4.4 and 4.7 eV. However, for our purposes, it is sufficient to consider an approximate value. In Table 24 of Ref.[69] the quantity $H_h = 40000$ cm^{-1} = 4.96 eV represents this constant of Eq. (31).

Latimer[175] estimated absolute hydration energies of ions. If an ion with the charge z and radius r is transferred from vacuo to an absolute dielectric, the energy gained is $-z^2/2r$ in atomic units, or $-z^2 \cdot 7.2$ eV/(r in Å). The reciprocal dielectric constant 0.013 of water is so small that it may be neglected in such considerations. The values given by Latimer[175] and later extended in Table 24 of Ref.[69] show that halide anions have approximately the hydration energy given by this dielectric theory, but that about 0.85 Å should be added to the conventional ionic radii of cations. This observation can be interpreted in different ways. One may argue that water cannot penetrate as close to the surface of the small cations, reminding one about the radii obtained by applying Stokes' law to molar conductivities. Another argument involve *local dielectric constants* introduced by Niels Bjerrum 1923 to explain $(pK_1 - pK_2 - 0.6)$ of dicarboxylic acids as a function of the distance between the two protons. Schwarzenbach[176, 177] evaluates reciprocal local dielectric constants (increased toward 0.3 for r decreasing to 4 Å) from such electrostatic contributions to equilibrium constants. Latimer's result would need this reciprocal value to be close to a-half. However, one should not conclude that all the interaction is electrostatic. Expressions of the type $-z^2 \cdot 7.2 \text{ eV}/(r + 0.85 \text{ Å})$ are so huge, that minor contributions do not matter too much, and the hydration energy is the result of all the changes, including covalent bonding, of the transfer of M^{+z} from vacuo to the aqueous solution. Actually, Scrocco and Salvetti[178] pointed out that the hydration energy of Ag^+ and Hg^{+2} are larger than expected from Latimer's theory. This does not prevent that we do not know the actual stereochemistry of these aqua ions.

The writer[4, 179] analyzed the *hydration energy differences* derived from Eq. (31) for aqua ions, where E^0 is known for the process increasing the oxidation number from $(z - 1)$ to z. It turns out that each transition group has a characteristic constant κ such that

$$I_{\text{chem}} = (2z - 1)\kappa \qquad (32)$$

with $\kappa = 5.3$ eV for the $3d$ group, 4.3 eV for the $4f$ group and 3.9 eV for the $5f$ group. Lower limits of $\kappa = 5.3$ eV can be obtained for beryllium and aluminium. It is clear that κ decreases with increasing ionic radii of the group, explaining the fact that I_4 of gaseous Ti^{+3} 43.27 eV is considerably larger than 36.72 eV of Ce^{+3} but that I_{chem} of titanium(III) aqua ions close to 4.4 eV is smaller than 6.2 eV of cerium(III) aqua ions. Actually, Eq. (32) is compatible with a hydration energy $-z^2 \kappa$ not taking the differential variation of ionic radii as a function of z into account. Thus, the effective ionic radius in the $3d$ group would be $7.2/5.3 = 1.36$ Å corresponding to the Latimer correction added to $r = 0.51$ Å. This result is an indirect confirmation of the order of magnitude of the quantity 4.5 eV of Eq. (31) because Eq. (32) would not obtain if I_{chem} were changed by a large constant. Though Eq. (32) is rather crude for chemical purposes, it represents the necessary link between I_z of gaseous M^{+z-1} (measured or calculated from approximate Ψ) and chemistry in aqueous solution. Thus, the oxidation states expected for unknown elements with Z above 102 can be predicted, at least for M(II) to M(IV), by choosing a reasonable value of κ and applying the condition that the ionization energy I_{chem} should be above 4.5 eV (otherwise, the aqua ion is thermodynamically unstable toward evolution of H_2 in 1 M acid) and the electron affinity equal (since adiabatic ionizations are considered) to I_{chem} of the aqua ion $(z - 1)$ should be below 5.73 eV in order not to evolve O_2 [67, 180]. For comparison, it may be noted that $\kappa = 5.4$ eV for F^-, 3.7 eV for Cl^- and 3.2 eV for I^-. The high hydration energy for fluoride may be connected not only with the small ionic radius but also the strong hydrogen bonding to water. The heat of evaporation of water is 0.5 eV/mole corresponding to the breaking of between one and two moderately strong hydrogen bonds. However, the moderate pK = 3 of HF in aqueous solution shows that HF must be considerably more stabilized, at least by entropy effects like $Al(H_2O)_5F^{+2}$.

B. Ahrland's Softness Parameter

Ahrland[29, 181] proposes that (I_{chem}/z) for the reaction Eq (30) is a parameter measuring the softness of definite central atoms. Table 3 gives values for $\sigma_A = (I_{chem}/z)$ which is a kind of energy *per electron* like standard oxidation potentials E^0. Presumably by accident, the lowest value is zero for lithium(I) and the highest well-established value is 5.1 eV for mercury(II) using the data of Rosseinsky[174] whereas Ahrland[29] suggested 4.6 eV. Ahrland has revised the value for the hydration energy of Ag^+ producing $\sigma_A = 4.1$ eV for silver(I). The value $\sigma_A = 1.5$ eV for copper(I) obtained from Table 24 in Ref.[69] looks anomalous and is not included in Table 3.

There is no obvious reason why σ_A should express Pearson's scale of softness. Nevertheless, it is evident that it succeeds much better than for instance the molar polarizabilities α given in Table 2. The border-line cases between hard and soft central atoms have σ_A around 3 eV, whereas typical hard behaviour is found when σ_A is below 2 eV. A mild criticism is that σ_A has a tendency to increase more with the oxidation number z than appropriate for the chemical softness, producing $\sigma_A = 3.1$ eV for iron(III) like for copper(II) and zinc(II). No aqua ions are known for $z = 0$ but an argument reducing σ_A *ad absurdam* for such a case is that it would automatically be zero, or more exactly the slightly negative hydration energy ΔH_{hydr} expected for a neutral atom.

The positive value of σ_A means that it costs energy to transform an atom of a metallic element (if the heat of sublimation of the condensed phase is added, an additional energy of usually 2 to 5 eV is needed, with exception of the low value 0.6 eV for mercury) to an aqua ion. This can be stated in a very provocative way[69]: the heats of formation of positively charged aqua ions (with the marginal exception of Li^+) are *positive* and not negative like conventional bond energies. Now, why are Na^+ and Ca^{+2} so stable in solution? This question (first asked by the contemporaries of Svante Arrhenius) can be answered by pointing out that a system does not contain free electrons with *zero* energy but that the loosest bound electron have a positive I called the *work-function* for metallic samples. Photo-electron spectra show I values between 8 and 15 eV for most gaseous and solid compounds, the lowest value[182] being 5.0 eV for the chromium(0) mesitylene complex $Cr(C_6H_3(CH_3)_3)_2$ having $N = 12$. It may be looking like putting the cart before the horses, but one may argue that the "driving force" forming aqua ions is the electron affinity representing a kind of Fermi level of condensed matter. The adiabatic electron affinity of water is 4.5 eV in the conditions of the hydrogen electrode. Said in other words, a calcium atom gains some $2(4.5 - \sigma_A) \sim 7.6$ eV by dissolving as Ca^{+2} in a solution having pH = 0 and producing 1 atm. H_2. Under equal circumstances, a zinc atom gains $\sim 2(4.5 - 3.1) = 2.8$ eV and a mercury atom *looses* 1.2 eV, in agreement with the fact that mercury vapour does not react with water.

Table 3. Softness parameters according to Ahrland and to Klopman, the sum of the z first ionization energies of the gaseous atom, and the enthalpies and free energies of hydration of M^{+z} according to Rosseinsky, and finally the heats of atomization of the elements, all in the unit 1 eV (= 23.05 kcal/mole)

	σ_A	σ_K	ΣI_n	$-\Delta H_{hydr}$	$-\Delta G_{hydr}$	ΔH_{atom}
H(I)	2.3	(−0.42)	13.60	11.3	11.3	2.26
Li(I)	0.0	−0.49	5.39	5.4	5.3	1.61
Be(II)	1.2	−3.75	27.53	25.1	−	3.33
Na(I)	0.9	0.00	5.14	4.2	4.3	1.08
Mg(II)	1.4	−2.42	22.68	19.9	19.8	1.66
Al(III)	1.6	−6.01	53.26	48.3	47.8	3.25
K(I)	1.0	−	4.34	3.3	3.5	0.93
Ca(II)	0.7	−2.33	17.98	16.5	16.5	2.00
Sc(III)	1.2	−	44.10	40.6	40.1	4.0
Cr(II)	2.1	−0.91	23.27	19.1	−	3.45
Mn(II)	2.0	−	23.07	19.1	19.0	2.97
Fe(II)	2.0	−0.69	24.05	20.0	19.8	4.20
Fe(III)	3.1	−2.22	54.70	45.4	44.9	4.20
Co(II)	1.8	−	24.92	21.3	20.8	4.56
Ni(II)	2.0	−0.29	25.80	21.7	21.4	4.40
Cu(II)	3.1	+0.55	28.02	21.8	21.6	3.53
Zn(II)	3.1	−	27.36	21.2	21.0	1.35
Ga(III)	2.9	−1.45	57.22	48.5	48.0	2.87
Rb(I)	1.0	−	4.18	3.2	3.3	0.89
Sr(II)	0.8	−2.21	16.72	15.0	15.0	1.70
Y(III)	0.6	−	39.14	37.5	37.3	4.47
Ag(I)	4.1	+2.82	7.58	(4.9)	(5.0)	3.00
Cd(II)	3.6	+2.04	25.90	18.7	18.7	1.18
In(III)	3.3	−	52.68	42.6	42.2	2.53
Cs(I)	1.2	−	3.89	2.7	2.9	−
Ba(II)	0.9	−1.89	15.22	13.5	13.7	1.82
La(III)	0.6	−4.51	35.81	34.0	−	3.8
Ce(IV)	2.0	−	73.24	65.1	−	−
Hg(II)	5.1	+4.64	29.19	19.0	19.0	0.63
Tl(I)	2.7	+1.88	6.11	3.4	3.6	1.93
Tl(III)	4.3	+3.37	56.37	43.4	−	1.93
Pb(II)	3.6	−	22.45	15.3	15.5	2.01

Whereas LiCl and CsF with disparate ionic radii dissolve exothermally in water, most halides of the alkaline metals dissolve without much heat evolution. This is a remarkable observation in view of the huge Madelung energy $-1.74/(r_M + r_x)$ in atomic units (14.3 eV/Å) relative to gaseous M^+ and X^- and it was first pointed out by Fajans[183] that this is compatible with the approximate equality

$$1.74/(r_M + r_x) \sim (1/2\,r_M) + (1/2\,r_x) \tag{33}$$

if the radii in the dielectric approximation to the right are slightly larger, as later treated quantitatively by Latimer[175]. However, Fajans extracted a quantity equivalent to the 4.5 eV of Eq. (31). It is interesting that the hydration "pays back" so accurately the enormous Madelung energy of the crystalline halides. This is a parallel to the general tendency discussed by Rabinowitch and Thilo[184] that Madelung potentials almost exactly compensate the ionization energies of the isolated atoms. Around 1930, this tendency was illustrated by many numerical calculations, whereas Eq. (32) for hydration energy has κ closely related to the coefficient $(a_1/2)$ of the *differential ionization energy* [4, 50].

When $\log \beta_4$ for $Hg(CN)_4^{-2}$ is 41.5 or when ΔH[30] is 63 kcal/mole the covalent bonding "pays back" 2.8 eV of the 2 σ_A lacking in the hydration of the aqua ion. This expression (10.2 eV) might be somewhat more extensively compensated in $Hg(CH_3)_2$ but seen from this point of view, σ_A expresses a deficiency of stabilization of the aqua ion relative to the gaseous neutral atom, and σ_A indicates the soft behaviour because this deficiency can be somewhat mitigated by strong covalent bonding. Said in other words, Li^+ is as stable in aqueous solution as the isolated lithium atom, whereas Hg^{+2} represents the opposite extreme. σ_A has many of the properties of a standard oxidation potential, and would be x eV lower if the quantity 4.5 eV was x higher. Thus, σ_A might have been introduced with the classical E^0 varying from -4.5 eV for lithium(I) to $+0.6$ eV for mercury(II) formed from gaseous atoms.

It is even possible to calculate σ_A exclusively from the classical E^0 for the oxidation of the metallic element to the aqua ion and from the heat of atomization ΔH_{atom} of the element given in Table 3. Neglecting for a moment the difference between ΔH and ΔG, it is a matter of constant energy that

$$-\Delta H_{hydr} = \Delta H_{atom} + \Sigma I_n - z(E^0 + 4.5)$$
$$z\sigma_A = \Sigma I_n + \Delta H_{hydr} = -\Delta H_{atom} + z(E^0 + 4.5) \tag{34}$$

As a numerical example, we may consider magnesium with $\Delta H_{atom} = 1.66$ eV and $\Sigma I_n = 22.68$ eV. Then, $E^0 = -2.38$ V corresponds to $\sigma_A = 1.29$ not far from the more sophisticated value 1.4 in Table 3. It may be noted that one does not need to calculate $-\Delta H_{hydr} = 20.1$ eV from Eq. (34) for this purpose. It is possible to use this approximation for obtaining σ_A for aqua ions not discussed by Rosseinsky[174]. For instance, the National Bureau of Standards Thermodynamical Tables give for Pd $\Delta H_{atom} = 93$ kcal/mole and $\Delta G_{atom} = 84$ kcal/mole. In this case, the sublimation heat of metallic palladium can be identified with the heat of atomization, because the closed-shell atom with the electron configuration $[Kr]4d^{10}$ presumably does not form a perceptible amount of diatomic or oligo-atomic molecules. Since half the ΔH for atomization of a hydrogen molecule is given as 2.26 eV in these tables, and half ΔG as 2.11 eV (corresponding to the parameters 4.56 and 4.41 eV of I_{chem}), the two values (ΔH and ΔG) combined with $E^0 = 0.99$ V of the standard oxidation potential of the metal to $Pd(H_2O)_4^{+2}$ both give $\sigma_A = 3.5$ for palladium(II), the same value as for cadmium(II) (where $-\frac{1}{2}\Delta H_{atom} + E^0 + 4.5 = -0.6 - 0.4 + 4.5 = 3.5$) and considerably smaller than for mercury (= $-0.3 + 0.85 + 4.5 = 5.05$). σ_A for copper(I) is only 1.55 and silver(I) 2.35 according to Eq. (34) whereas it is 2.6 for chromium(III), 3.4 for cobalt(III) and 4.1 for bismuth(III).

It is important to note that (contrary to the discussion of hydration energy) the differences between various σ_A are not influenced by the knowledge of the parameter

55

of I_{chem} in Eq. (31), close to 4.5 eV, and Eq. (34) is closely related to the ideas of Abegg and Bodländer except that E^0 are diminuished by $(\Delta H_{atom}/z)$ which is a much larger quantity for metals such as Fe, Ru, Rh, W, Re, Ir and Pt than for Hg and Tl.

C. Klopman's Approach

Klopman[185] previously applied a semi-empirical quantum-chemical treatment to organic molecules with the main purpose of predicting bond energies and heats of formation. The values obtained were remarkably accurate, usually between 2 and 3 kcal/mole (0.1 eV). After the symposium at the Cyanamid European Research Institute 1965, Klopman[186] proposed a set of softness parameters derived from a theory of ionic solvation taking into account the variation of the orbital energy[50] with its occupation number. For the aqua ion of M^{+z}, Klopman proposes the orbital energy being $0.75\,I_{z-1} + 0.25\,I_{z-2}$ of the gaseous entities, as a compromise between the "charge-controlled reaction" with little charge-transfer (where the electron affinity of the loosest bound orbital I_{z-1} is decisive) and the "frontier-controlled reaction" with strong covalency of the frontier orbitals (where the mean value $(I_{z-1} + I_{z-2})/2$ is appropriate). Klopman[186] derives a rather complicated expression for the *desolvation energy* which turns out to be very close to $K_{Gilles}/(r_{ion} + 0.82\,\text{Å})$ where the parameter K_{Gilles} is 7.1 eV for M(I), 23.1 for M(II), 42.7 eV for M(III) and 65.7 eV for M(IV). In Table 3 are given the value in eV of $\sigma_K = -E_n^{\ddagger}$ where the latter quantity was defined by Klopman as the difference between the desolvation energy and the orbital energy. For instance, barium has $I_1 = 5.21$ eV and $I_2 = 10.0$ eV and hence, the orbital energy 8.80 eV. Since $r_{ion} = 1.34$ Å, the desolvation energy is $23.1/2.16 = 10.69$ eV. The reason why we have changed the sign of $\sigma_K = 8.80 - 10.69 = -1.89$ eV for Ba(II) is that it is nicer to have a softness parameter increasing with the softness.

It may be noted that $\sigma_K = -0.42$ eV for hydrogen(I) is derived from the experimental desolvation energy of the proton assumed to be 10.8 eV rather than 11.3 eV in Table 3. If one calculated $(K_{Gilles}/0.82\,\text{Å}) = 8.9$ eV one would obtain $\sigma_K = +1.48$ eV which might be defended too from the strong affinity of protons to H^-, CN^- and CH_3^-. It may also be noted that Klopman[186] chooses to consider ionization of $4s$ orbitals in Cr(II), Ni(II) and Cu(II) rather than $3d$.

There is no doubt that Ahrland's approach is more transparent. The main differences are the use of the sum of all $I_n(n = 1$ to $z)$ and of experimental hydration energies in the former case. An appealing aspect of Klopman's treatment is the desolvation energy only depending on ionic radii and charges. It is noted that K_{Gilles} is *roughly* proportional to $(2z - 1)$ since the model is based on differential charge transfer.

Both models can be applied to *anions*. Eq. (30) can be extended in an obvious way to give σ_A for halides (but not for H^- unstable in water). If one is willing to accept ionic radii for H^- and diatomic anions, σ_K can be evaluated[186] as the *sum* of the orbital energy (0.25 ionization energy +0.75 electron affinity) and the desolvation energy, $(7.1\ \text{ev}/r_{ion})$ both with opposite sign. σ_K vary more than σ_A for halides, but the variation of these negative values is less pronounced than for the cations.

	F^-	Cl^-	Br^-	I^-	H^-	OH^-	CN^-	SH^-	
σ_A	−8.7	−7.5	−6.9	−6.1	−	−	−	−	
r_{ion}	1.36	1.81	1.95	2.16	2.08	1.40	2.60	1.84	(35)
σ_K	−12.18	−10.45	−9.22	−8.31	−7.37	−10.45	−8.78	−8.59	

D. Ionization Energies of Valence Electrons and Partly Filled Shells

In the period 1955–70 the partly covalent bonding in "ligand field" theory was considered an extension of the Hückel treatment of heteronuclear molecules. The chromophore MX_N had diagonal elements of energy H_M *less negative* than H_X of the ligating atoms, and the eigen-values supposed to represent ionization energies I (with opposite sign) are evaluated for a secular determinant having as non-diagonal elements the product of the *overlap integral* S_{MX} multiplied by an average energy, for which Wolfsberg and Helmholz proposed $k(H_M + H_X)/2$ with k between 1.5 and 2. This model has the advantage of making the anti-bonding MO twice or thrice as anti-bonding MO are bonding, in agreement with experience. It was suggested[50] in 1962 that certain complexes of oxidizing central atoms and reducing ligands (such a $CuBr_4^{-2}$ and OsI_6^{-2}) might turn out to have higher I of the partly filled shell than of the loosest bound MO mainly situated on the ligand. The mechanism of this apparent paradox is the huge *difference between I and electron affinity* of the partly filled shell (for which the theoretical estimates have the order of magnitude 10 eV in the d groups and 20 eV in the $4f$ group) which is recognized[4,67,187] to play a decisive rôle in the chemical behaviour of the five transition groups, and in particular, in the choice of oxidation states when combined with the hydration or Madelung energies.

However, the *photo-electron spectra*[151,170] have shown the situation of higher I of the partly filled shell to be far more frequent than a few exceptional cases in the d group. A great surprise is that[188,189] most $4f$ group compounds (with exception of Ce(III), Pr(III) and Tb(III)) have comparable or *higher I* of the partly filled shell than of the $2p$ orbitals of fluoride or of oxygen-containing ligands. It is normally felt that the condition for strong covalent bonding is comparable H_M and H_X. Seen from the strict point of view of solving Schrödinger's equation, a very small S_{MX} cannot prevent covalent delocalization in the case of coinciding $H_M = H_X$ Nevertheless, it is one of the best established facts about $4f$ group complexes[20,50,190] that the covalent bonding, both as far goes radial expansion adapting to decreased fractional charge and delocalization on the ligands corresponding to formation of anti-bonding orbitals[191], is very weak, as seen from the nephelauxetic effect (decreasing phenomenological parameters of interelectronic repulsion) and spin-orbit coupling parameter ζ_{4f} varying less than one percent in different ytterbium(III) compounds. The corresponding problem of finding LCAO eigen-vectors is far more complicated than first believed and constitutes what is colloquially called "the third revolution in ligand field theory"[58,171,192]. It is interesting that the *angular overlap model* originally intended[191] to describe anti-bonding effects on the seven $4f$ orbitals proportional to S_{MX}^2 remains valid, as it does for d-group complexes[67,82,193] independently of this difficulty, and it can be shown[67,193] to be equi-consequential with a singular contact potential acting in the vicinity of each ligand nucleus.

In some cases, the d-like orbitals have lower I than the MO mainly localized on the ligands. The first known case was VCl_4 having one $3d$-like electron with $I = 9.4$ eV well below the filled MO starting at 11.8 eV like in the analogous $TiCl_4$ representing linear combinations of essentially chlorine $3p$ orbitals. Such evidence is usually known for gaseous molecules[151] where the good resolution allows vibrational structure to be detected, and a typical case is all three $M(CO)_6$ (M = Cr, Mo and W) having $I = 8.5$ eV of the six d-like electrons (in the following, we give the vertical I corresponding to the maxima (or the baricenters of vibrational structures) of photo-electron signals). Nixon[194] studied molecules such as $Cr(PF_3)_6$, $Fe(PF_3)_5$ and $Rh(PF_3)_4H$ to which "ligand field" (and to a large extent, even angular overlap) arguments can be applied. For whatever reason, the d-like electrons have I one to two eV higher in PF_3 than in the corresponding CO complexes. One may think of back-bonding to empty orbitals increasing I, either by stabilizing the d-like orbitals being the lower eigen-values, or indirectly by increased fractional charge of the central atom. However, it has become clear[58,192] that the minimization of the total energy of the molecule does not correspond to a necessary minimum of the energy of the partly filled shell. Partly covalent bonding pulls in direction of *lower I* of d-like electrons by the predominant influence of decreased fractional charge of the central atom also, to a large extent, removing the intrinsic difference between cobalt(II) and (III) or even nickel(II) and (IV). These central atoms have far lower $I(3d)$ than gallium(III) close to 26 eV[58]. The recent results from photo-electron spectra have explained why the partly filled shell in most cases remain fairly localized on the central atom[4,67] not so much because of a much lower electronegativity than of the ligands, but rather due to the large difference between I and the electron affinity (also favouring a definite integer as the number of $4f$ electrons in the metallic elements, alloys and compounds). Even the rather unique exception TmTe, a stoichiometric metal crystallizing in NaCl type, contains about half $4f^{13}$ Tm(II) and half $4f^{12}$ Tm(III) on an instantaneous picture, but the photo-electron spectrum[195] is an unambiguous superposition of the two sets of signals.

For our purposes, the results of photo-electron spectra of solids[58,189,196] are somewhat worrying in sofar the ten $4d$-like electrons of silver(I) have almost the same I as the $5p$-like electrons of iodide, whereas the $4d$ electrons of cadmium(II) have I close to 17 and indium(III) about 24 eV. For a conventional LCAO description, this evidence clearly indicates that the filled d shell hardly is influenced by chemical bonding, and that the obvious chemical differences between Cd(II) and Sr(II) or between In(III) and Y(III) are due to covalent bonding of the empty $5s$ orbital (the $4d$ and $5s$ orbitals cannot mix in a one-electron description if the chromophore remains cubic on an instantaneous picture with three equivalent Cartesian axes) or possibly to the $5p$ orbitals, besides the smaller ionic radii of Cd(II) and In(III). However, the situation is far worse in solid silver(I) halides. It has been known for half a century[184] that the Madelung energy overestimates the heat of formation of AgF from gaseous Ag^+ and F^- by a few percent[4] like it does for the isotypic alkali halides. This discrepancy was explained by Born, Haber and Fajans[183] as the destabilization due to the repulsion between the closed cores. That such a destabilization with a dramatic increase for shorter internuclear distances *must* occur can be seen from the fact that the Madelung energy alone gets x times more negative if all the distances are decreased by the same factor x. Solid AgCl and AgBr are somewhat exceptional by having ΔH from gaseous ions a few percent higher[4] than predicted by the Madelung theory. This has always been taken as evidence for covalent bonding[184] but the problem now is that the coinciding I of silver $4d$ and halide p electrons produces a destabilization of the same kind as core-core repulsion preventing solids from imploding. Hence, AgF avoiding this coincidence should be more stable. This is the reason why *continuum effects*[67] were invoked to explain the excess stability of AgCl and AgBr by deformation of the constituent orbitals (having no direct effect on the photo-electron spectra). These materials are also unusual by forming the *latent image* known from photography, and where it has been suggested[197] that two additional electrons are delocalized on a cluster of twelve silver atoms.

The I values being one-electron quantities show a contrast to valence-bond hybridization. Even CH_4 has two signals centered around 14 and 23 eV to be compared with $Ne\,2p$ 21.6 and $Ne\,2s$ 48.4 eV and the intermediate ten-electron system H_2O 12.6, 14.8, 18.6 and 32.2 eV which could be interpreted as a "ligand field" in the point-group C_{2v} separating the three $2p$ orbitals of oxygen($-$II) remaining rather separate from $2s$. It is difficult to tell whether hybridization occurs or not in the groundstate, but its importance has been overestimated[48,68].

E. Ionization Energies of Inner Shells and Relaxation Effects

Around 1968, it was believed[198] that the chemical shift dI of ionization energies of inner shells depends mainly on the oxidation state of the atom considered. Elements able to change their oxidation number by 8 units, such as nitrogen, sulphur and chlorine, vary indeed their I values some 12 eV. However, this is to a large extent a question of the ligating atoms. Thus, $I(S\,2p)$ of SF_6 is 180.4 eV to be compared with the interval 175 to 173 eV for various sulphates and 170.2 eV for H_2S[152]. When Berthou and the writer[189] measured photo-electron spectra of 600 nonmetallic compounds containing 77 elements, distinct exceptions were found. Thus, Tl_2O_3 has *lower I* than the 21 thallium(I) compounds measured, PbO_2 lower than all the 15 lead(II) compounds studied, and most cobalt(III) complexes lower I than Co(II). The general trends of high I in the fluoride of a given central atom and low I in the oxide (known for pronounced nephelauxetic effect) and sulphur-containing compounds are compatible with a dependence on the fractional atomic charge [4,39,67] though the iodides and cyanides show unexpectedly large I seen from this point of view. This variation (usually 2 to 5 eV for a given oxidation state) could be understood a s a first-order perturbation from the combined Hartree + Madelung potential not contributing[50] to the total energy of the system, and explaining why X-ray absorption spectra show chemical shifts an order of magnitude larger than the X-ray emission lines due to transitions between two shells perturbed by roughly the same Hartree + Madelung potential.

However, it is certain that this cannot be the whole truth. The *relaxation effect* consisting of an energy decrease by the adaptation of the electronic density of the other orbitals (contrary to the behaviour described by Koopmans) and a comparatively smaller energy increase due to a part of the correlation energy disappearing by the ionization, is known to be 20 eV for the $1s$ orbital of the neon atom having $I = 870.2$ eV. Quite generally, the *intra-atomic* part of the relaxation energy is empirically[171] close to 0.8 eV times the square-root of I (in eV) of inner shells. Careful calculations of the relaxation energy[199] in simple carbon-, nitrogen-, and oxygen-containing molecules are between 15 and 23 eV for $1s$ ionization. It is clear that for our purpose, the inter-atomic part of the relaxation energy are more interesting. Indeed, potassium(I) salts[170,189] show a variation of $I(K2p_{3/2})$ which can hardly be ascribed to changes of the fractional charge. Table 4 gives the values of I' corrected

Table 4. Ionization energies in eV of the ($j = 3/2$) component of the $2p$ shell of potassium obtained by photo-electron spectrometry, and corrected for the quasi-stationary positive potential of the non-conductors loosing electrons

KBF_4	299.3	KIO_3	298.1
KPF_6	299.2	KIO_4	298.0
K_3RhF_6	299.0	K_2HgI_4	298.0
K_2SiF_6	298.8	KF	297.9
K_2GeF_6	298.6	KI	297.9
K_4BiI_7	298.6	KCl	297.8
K_2BeF_4	298.5	K_2HOAsO_3	297.8
K_2NbF_7	298.5	K_2UF_6	297.8
K_2HfF_6	298.5	$KNiF_3$	297.7
K_2TiF_6	298.4	$KNOsO_3$	297.7
K_2PtCl_6	298.3	$K_2Pt(SeCN)_6$	297.7
KBr	298.3	$KAg(CN)_2$	297.5
K_2TaF_7	298.2	$KBrO_3$	297.3
		$KSeCN$	296.5

for the effect of charging of non-conducting samples acquiring a quasi-stationary positive potential in the photo-electron spectrometer. When this potential is above 1 V (it reaches 5 V for certain anhydrous fluorides) we argue that it can be measured as the distance between two carbon $1s$ signals seen when the powdered sample is distributed[200] on the adhesive side of one-sided scotch tape (such as 600 P from the 3M Company; other brands are not always an aliphatic hydrocarbon but contains poly-ether or C-Cl bonds). Since gaseous hydrocarbons have $I(C\ 1s)$ between 290.8 and 290.4 eV[152] of carbon atoms exclusively surrounded by hydrogen and other carbon atoms, and since the mild dependence on molecular size suggests an approximate saturation of the relaxation effects, we argue[201] that these signals indicate $I = 290.0$ eV relative to *vacuo* (whereas most authors would write $I^* = 285$ eV relative to a Fermi level of the spectrometer) and we define C'_{st} as the difference between 290 eV and the highest of the two $I^*(C\ 1s)$ values recorded by the Varian IEE-15 apparatus. Then, we define $I' = I^* + C'_{st}$ for the signals of the other elements. This definition gives good agreement with independent evidence for the valence electrons of solids and in gaseous molecules. In the case of conducting samples, one sees only one carbon $1s$ signal at $I^* = 290 - C_{st}$ and $I = I^* + C_{st}$. Not only metals in electric contact with the instrument, but also certain samples with strong hydrogen bonds (say, salt hydrates) or containing conjugated constituents (such as tetraphenyl-borates or salts of tetraphenylarsonium or methylene blue cations) are sufficiently conducting under the X-ray bombardment (with 1253.6 or 1483.6 eV photons) to fall in the latter category.

It is quite striking that the fluorine-containing anions in Table 4 show I' ($K2p_{3/2}$) in average 298.7 eV to be compared with 296.5 eV for the soft selenocyanate and 297.7 eV for $K_2Pt(SeCN)_6$. However, Table 4 is incomplete for the large anions, there is no doubt[170,189] that K_2PtI_6, K_2PdCl_4, $K_2Pd(CN)_4$, $K_2Ni(CN)_4$, $KAu(CN)_2$, K_2PtBr_6, K_2IrBr_6, $K_2Hg(SCN)_4$, $K_4Fe(CN)_6$, $K_3Cr(NCS)_6$, $KMnO_4$, $K_2Pt(SCN)_6$ and $K_3Fe(CN)_6$ also have low ionization energies, but the distance between the two C $1s$ signals indicating the quasi-stationary potential (which we believe is the distance from the lower limit of the empty conduction band of the sample to the Fermi level of the metallic surroundings) cannot be evaluated.

Actually, the results in Table 4 are even more surprising if corrected for the Madelung potential. If $V_{Mad} = 9.4$ eV in KF, 8.0 KCl, 7.6 KBr and 7.1 eV in KI are added to the I' values, one obtains 307.3, 305.8, 305.9 and 305.0 eV, respectively. The fluoride-containing anions have more than twice as large radii, and there are good reasons to believe that their I' values corrected for the Madelung potential would be lower, probably close to 304 eV. This idea[171] of more important relaxation effects in fluorine-containing anions is corroborated by the lower I' observed in K_2UF_6 and the perovskite $KNiF_3$ not containing monomeric anions but an extended lattice with fluoride bridges. That the large pseudohalogen- and halogen-containing anions of noble central atoms produce low I' values is accentuated by the weaker Madelung potential. A rather extreme case is $K^+B(C_6H_5)_4^-$ where $I(K2p_{3/2}) =$ 298.9 eV is an upper limit to I' (a plausible I' would be 298 eV) and the Madelung potential is at most 2 eV. Hence, it seems that the inter-atomic relaxation is 7 eV larger in the tetraphenylborate than in the fluoride.

Seen from the point of view of the chemist, the relaxation effects remaining in $I'(K2p)$ after correction for the Madelung potential represent the propensity of the neighbour atom to establish covalent bonding to the K^{+2} lacking a $2p$ electron within 10^{-15} sec. It would be a more transparent question to consider the groundstate of potassium(II) lacking a $3p$ electron, and actually, the $I'(K3p)$ varying between 23 and 21 eV (to be compared with 15.8 eV for argon $3p$) show the same trends, but the photo-electron signals are much weaker[202, 203] than $K2p$. It is not surprising that the ephemeric super-oxidant K(II) shows many of the characteristics of soft central atoms such as Pd(II) and Au(III).

It is conceivable that one may extract other evidence from photo-electron spectra of relevance for the physical significance of soft and hard behaviour. Thus, it is characteristic[204] that the post-transitional atoms Cd(II), In(III) and Hg(II) show much smaller I' differences between fluorides and oxides than other elements in the oxidation state z where the order of magnitude of dI is z eV with the insoluble iodates occupying a position intermediate between fluorides and oxides. Noble gases[205] embedded in metallic copper, silver or gold show inter-atomic relaxation effects[171] amounting to 2 to 3 eV. In all of these cases, phenomena with a time-scale around 10^{-17} sec obey the Manne-Åberg principle[206] of electronic relaxation in the primary signals, in addition to the analogous Franck-Condon principle of nuclear positions. This may be much closer to what chemists loosely name polarizability than the α values of Table 2. Chemistry is intrinsically a question of the influence of the adjacent atoms.

VI. References

1) Werner, A.: Neuere Anschauungen auf dem Gebiete der anorganischen Chemie (3. Ed.). Braunschweig: Vieweg 1913.
2) Yatsimirski, K. B., Vasilev, V. P.: Instability Constants of Complex Compounds. Oxford: Pergamon Press 1960.
3) "Stability Constants". Chemical Society Special Publications No. 6, 7, 17 and 25. London: 1957, 1958, 1964 and 1971.
4) Jørgensen, C. K.: Oxidation Numbers and Oxidation States. Berlin–Heidelberg–New York: Springer 1969.
5) Bjerrum, J.: Mat. fys. Medd. Danske Vid. Selskab *11*, no. 5 and 10 (1932); ibid. *12*, no. 15 (1934).
6) Bjerrum, J.: Metal Ammine Formation in Aqueous Solution. Copenhagen: Haase and Son; 1941 (2. ed. 1957) and Chem. Rev. *46*, 381 (1950).
7) Krishnamurthy, R., Schaap, W. B., Perumareddi, J. R.: Inorg. Chem. *6*, 1338 (1967).
8) Wolsey, W. C., Reynolds, C. A., Kleinberg, J.: Inorg. Chem. *2*, 463 (1963).
9) Jørgensen, C. K., Preetz, W.: Z. Naturforsch. *22a*, 945 (1967).
10) Jørgensen, C. K., Preetz, W., Homborg, H.: Inorg. Chim. Acta *5*, 223 (1971).
11) Martell, A. E., Calvin, M.: Chemistry of the Metal Chelate Compounds. New York: Prentice-Hall 1952.
12) Lewis, J., Wilkins, R. G.: Modern Coordination Chemistry. New York: Interscience 1960.
13) Rossotti, F. J. C., Rossotti, H.: The Determination of Stability Constants. New York: McGraw-Hill 1961.
14) Schläfer, H. L.: Komplexbildung in Lösung. Methoden zur Bestimmung der Zusammensetzung und der Stabilitätskonstanten gelösten Komplexverbindungen. Berlin–Heidelberg–New York: Springer 1961.
15) Beck, M. T.: Chemistry of Complex Equilibria. London: Van Nostrand-Reinhold 1970.
16) Högfeldt, E. (ed.): Coordination Chemistry in Solution – In Memory of Lars Gunnar Sillén. Stockholm: Transactions of the Royal Institute of Technology 1972.
17) Burger, K.: Coordination Chemistry, Experimental Methods. Cleveland: Chemical Rubber Company 1973.
18) Pearson, R. G.: Chem. Comm. (London) *1968*, 65.
19) Müller, A., Diemann, E., Jørgensen, C. K.: Structure and Bonding *14*, 23 (1973).
20) Jørgensen, C. K., Rittershaus, E.: Mat. fys. Medd. Danske Vid. Selskab *35*, no. 15 (1967).
21) Müller, E.: Origin and Distribution of the Elements (ed.: L. H. Ahrens) 155. Oxford: Pergamon Press 1968.
22) Prue, J. E. Schwarzenbach, G.: Helv. Chim. Acta *33*, 974 and 985 (1950).
23) Jørgensen, C. K.: Acta Chem. Scand. *10*, 887 (1956).
24) Schwarzenbach, G.: Experientia Suppl. *5*, 162 (1956).
25) Schwarzenbach, G.: Adv. Inorg. Radiochem. *3*, 257 (1961).
26) Jørgensen, C. K.: J. Inorg. Nucl. Chem. *24*, 1571 (1962).
27) Jørgensen, C. K.: Inorg. Chim. Acta Rev. *2*, 65 (1968).
28) Ahrland, S., Chatt, J., Davies, N. R.: Quart. Rev. (London) *12*, 265 (1958).
29) Ahrland, S.: Structure and Bonding *5*, 118 (1968).

30) Ahrland, S.: Structure and Bonding *15*, 167 (1973).
31) Poulsen, I., Bjerrum, J.: Acta Chem. Scand. *9*, 1407 (1955).
32) Pearson, R. G.: J. Amer. Chem. Soc. *85*, 3533 (1963).
33) Bjerrum, J.: Naturwiss. *38*, 461 (1951).
34) Pearson, R. G.: J. Chem. Educ. *45*, 581 and 643 (1968).
35) Pearson, R. G. (ed.): Hard and Soft Acids and Bases. Stroudsburg, Penn.: Dowden, Hutchinson and Ross 1973.
36) Jørgensen, C. K.: Structure and Bonding *1*, 234 (1966).
37) Williams, R. J. P., Hale, J. D.: Structure and Bonding *1*, 249 (1966).
38) Fenske, R. F., Caulton, K. G. Radtke, D. D., Sweeney, C. C.: Inorg. Chem. *5*, 951 and 960 (1966).
39) Jørgensen, C. K.: Helv. Chim. Acta Fasc. extraord. Alfred Werner 131 (1967).
40) Jørgensen, C. K.: Inorganic Complexes. London: Academic Press 1963.
41) Christiansen, K. A., Høeg, H., Michelsen, K., Bech Nielsen, G., Nord, H.: Acta Chem. Scand. *15*, 300 (1961).
42) Orgel, L. E.: Introduction to Transition-Metal Chemistry. London: Methuen 1960 (2. ed. 1966).
43) Beurskens, P. T., Cras, J. A., Steggerda, J. J.: Inorg. Chem. *7*, 810 (1968).
44) Warren, L. F., Bennett, M. A.: J. Amer. Chem. Soc. *96*, 3340 (1974).
45) Jørgensen, C. K.: Progress Inorg. Chem. *12*, 101 (1970).
46) Jørgensen, C. K.: Progress Inorg. Chem. *4*, 73 (1962).
47) Jørgensen, C. K., Horner, S. M. Hatfield, W. E., Tyree, S. Y.: Int. J. Quantum Chem. *1*,191 (1967).
48) Jørgensen, C. K.: Chimia (Aarau) *25*, 109 (1971).
49) Jørgensen, C. K.: Chem. Phys. Letters *27*, 305 (1974).
50) Jørgensen, C. K.: Orbitals in Atoms and Molecules. London: Academic Press 1962.
51) Valentine, J. S.: Chem. Comm. (London) 857 (1973).
52) Jørgensen, C. K.: Chimia (Aarau) *28*, 605 (1974).
53) Bjerrum, J.: Fysisk Tidsskrift (Copenhagen) *71*, 113 (1973).
54) Crouch, E. C. C., Pratt, J. M.: Chem. Comm. (London) *1969*, 1243.
55) Kruck, T.: Angew. Chem. *79*, 27 (1967).
56) Dasent, W. E.: Non-existent Compounds. New York: Marcel Dekker 1965.
57) Jørgensen, C. K.: Inorg. Chem. *3*, 1201 (1964).
58) Jørgensen, C. K.: Chimia (Aarau) *28*, 6 (1974).
59) Siebert, H.: Z. anorg. Chem. *327*, 63 (1964).
60) Pratt, J. M. Thorp, R. G.: J. Chem. Soc. (A) 187 (1966).
61) Iwata, M., Saito, Y.: Acta Cryst. *B29*, 822 (1973).
62) Malatesta, L.: Progress Inorg. Chem. *1*, 284 (1959).
63) Schmidtke, H. H., Garthoff, D.: Z. Naturforsch. *24a*, 126 (1969).
64) Haim, A., Sutin, N.: J. Amer. Chem. Soc. *88*, 434 (1966).
65) Wasson, J. R., Trapp, C.: J. Inorg. Nucl. Chem. *30*, 2437 (1968).
66) Burmeister, J. L.: Coord. Chem. Rev. *1*, 205 (1966) and *3*, 225 (1968).
67) Jørgensen, C. K.: Modern Aspects of Ligand Field Theory. Amsterdam: North-Holland 1971
68) Jarvie, J., Wilson, W., Doolittle, J., Edminston, C.: J. Chem. Phys. *59*, 3020 (1973).
69) Jørgensen, C. K.: Absorption Spectra and Chemical Bonding in Complexes. Oxford: Pergamon Press 1962.
70) Reedijk, J., Van Leeuwen, P. W. N. M., Groeneveld, W. L.: Rec. trav. chim Pays-Bas *87*, 129 (1968).
71) Reinen, D.: Ber. Bunsenges. *69*, 82 (1965) and Theoret. Chim. Acta *5*, 312 (1966).
72) Burns, R. G.: Mineralogical Applications of Crystal Field Theory. Cambridge: University Press 1970.
73) Romano, V., Bjerrum, J.: Acta Chem. Scand. *24*, 1551 (1970).
74) Hathaway, B. J.: Structure and Bonding *14*, 49 (1973).
75) Jørgensen, C. K.: Acta Chem. Scand. *8*, 175 (1954).
76) Schneider, W., Baccini, P.: Helv. Chim. Acta *52*, 1955 (1969).
77) Smith, D. W.: Structure and Bonding *12*, 49 (1972).

78) Herzberg, G.: Electronic Spectra and Electronic Structure of Polyatomic Molecules. Princeton: Van Nostrand 1966.

79) Tanabe, Y., Sugano, S.: J. Phys. Soc. Japan 9, 753 and 766 (1954).

80) Fenske, R. F.: J. Amer. Chem. Soc. 89, 252 (1967).

81) Jørgensen, C. K.: Adv. Chem. Phys. 5, 33 (1963).

82) Schäffer, C. K., Jørgensen, C. K.: Mat. fys. Medd. Danske Vid Selskab 34, no. 13 (1965).

83) Shimura, Y., Tsuchida, R.: Bull. Chem. Soc. Japan 29, 311 (1956).

84) Reinen, D.: Z. Naturforsch. 23a, 521 (1968).

85) Jørgensen, C. K.: The Biochemistry of Copper (p. 1). New York: Academic Press 1966.

86) Sone, K., Utsono, S., Ogura, T.: J. Inorg. Nucl. Chem. 31, 117 (1969).

87) Ciampolini, M.: Structure and Bonding 6, 52 (1969).

88) Norgett, M. J., Thornley, J. H. M., Venanzi, L. M.: Coord. Chem. Rev. 2, 99 (1967).

89) Day, P. Jørgensen, C. K.: J. Chem. Soc. 6226 (1964).

90) Dunitz, J. D., Orgel, L. E.: Adv. Inorg. Radiochem. 2, 1 (1960).

91) Jørgensen, C. K.: Acta Chem. Scand. 10, 500 (1956).

92) Rasmussen, L., Jørgensen, C. K.: Acta Chem. Scand. 22, 2313 (1968).

93) Reinhardt, R. A., Coe, J. S.: Inorg. Chim. Acta 3, 438 (1969).

94) Elding, L. I.: Inorg. Chim. Acta 6, 647 and 683 (1972).

95) Chang, J. C., Bjerrum, J.: Acta Chem. Scand. 26, 815 (1972).

96) Reinhardt, R. A., Monk, W. W.: Inorg. Chem. 9, 2026 (1970).

97) Coe, J. S., Lyons, J. R., Hussain, M. D.: J. Chem. Soc. (A) 90 (1970).

98) Roulet, R., Ernst, R.: Helv. Chim. Acta 54, 2357 (1971).

99) DeBerry, W. J., Reinhardt, R. A.: Inorg. Chem. 11, 2401 (1972).

100) Ernst, R., Roulet, R.: Chimia (Aarau) 28, 347 (1974).

101) Srivastava, S. C., Newman, L.: Inorg. Chem. 11, 2855 (1972).

102) Rasmussen, L., Jørgensen, C. K.: Inorg. Chim. Acta 3, 543 (1969).

103) Schmidtke, H. H., Jørgensen, C. K.: Chem. Phys. Letters 5, 202 (1970).

104) Hewkin, D. J., Poë, A. J.: J. Chem. Soc. (A) 1884 (1967).

105) Goddard, J. B., Basolo, F.: Inorg. Chem. 7, 936 (1968).

106) Coe, J. S., Lyons, J. R.: J. Chem. Soc. (A) 2669 (1969).

107) Jørgensen, C. K.: Inorg. Chim. Acta 3, 313 (1969).

108) Livingstone, S. E.: J. Proc. Roy. Soc. New South Wales 85, 151 and 86, 32 (1952).

109) Josephsen, J., Schäffer, C. E.: Acta Chem. Scand. 23, 2206 (1969) and 24, 2929 (1970).

110) Sigel, H., Griesser, R., McCormick, D. B.: Inorg. Chim. Acta 7, 594 (1973).

111) Carty, A. J., Chieh, P. C.: Chem. Comm. (London) 158 (1972).

112) Rund, J. V.: Inorg. Chem. 7, 24 (1968).

113) Rasmussen, L., Jørgensen, C. K.: Inorg. Chim. Acta 3, 547 (1969).

114) Parthasarathy, V., unpublished results.

115) Kasahara, A.: Bull. Chem. Soc. Japan 41, 1272 (1968).

116) Schurter, M., unpublished results.

117) Cope, A. C., Friedrich, E. C.: J. Amer. Chem. Soc. 90, 909 (1968).

118) Parshall, G. W.: Accounts Chem. Res. 3, 139 (1970).

119) Trofimenko, S.: Inorg. Chem. 12, 1215 (1973).

120) Braunstein, P., Dehand, J., Pfeffer, M.: Inorg. Nucl. Chem. Letters 10, 581 (1974).

121) Arzoumanidis, G., Rauch, F. C.: Chem. Comm. (London) 666 (1973).

122) Harris, S. J., Tobias, R. S.: Inorg. Chem. 8, 2259 (1969).

123) Mason, W. R., Gray, H. B.: Inorg. Chem. 7, 55 (1968).

124) Elding, L. I.: Acta Chem. Scand. 24, 1331, 1341 and 1527 (1970).

125) Griffith, W. P.: The Chemistry of the Rarer Platinum Metals (Os, Ru, Ir and Rh). London: Interscience 1967.

126) Hartley, F. R.: The Chemistry of Platinum and Palladium. London: Applied Science Publications 1973.

127) Day, P., Seal, R. H.: J. Chem. Soc. (Dalton) 1972, 2054.

128) Volbert, F.: Z. physik. Chem. A149, 382 (1930).

129) Jørgensen, C. K., Pouradier, J.: J. Chim. physique 67, 124 (1970).

130) Hancock, R. D., Finkelstein, N. P., Evers, A.: J. Inorg. Nucl. Chem. 36, 2539 (1974).

131) Jørgensen, C. K.: Energy Levels of Complexes and Gaseous Ions. Copenhagen: Gjellerup 1957.
132) Orgel, L. E.: J. Chem. Soc. 4186 (1958).
133) Cotton, F. A., Harris, C. B.: Inorg. Chem. 6, 369 (1967).
134) Meier, M.: Dissertation No. 3988, ETH Zürich (1967).
135) Wirth, T. H., Davidson, N.: J. Amer. Chem. Soc. 86, 4322 (1964).
136) Schwarzenbach, G., Schellenberg, M.: Helv. Chim. Acta 48, 28 (1965).
137) Nyholm, R. S., Vrieze, K.: J. Chem. Soc. 1965, 5331 and 5337.
138) Cuthforth, B. D., Gillespie, R. J., Ireland, P. R.: Chem. Comm. (London) 1973, 723.
139) Duffy, J. A., Ingram, M. D.: Chem. Comm. (London) 1971, 443 and Inorg. Chim. Acta 7, 594 (1973).
140) Walton, R. A., Matthews, R. W., Jørgensen, C. K.: Inorg. Chim. Acta 1, 355 (1967).
141) Gillespie, R. J.: Molecular Geometry. London: Van Nostrand 1972.
142) Orgel, L. E.: J. Chem. Soc. 1959, 3815.
143) Merritt, C., Hershenson, H. M., Rogers, L. B.: Analyt. Chem. 25, 572 (1953).
144) Couch, D. A., Wilkins, C. J., Rossman, G. R., Gray, H. B.: J. Amer. Chem. Soc. 92, 307 (1970).
145) Pearson, R. G.: J. Amer. Chem. Soc. 91, 4947 (1969).
146) Berggren, K. F.: J. Chem. Phys. 60, 3399 and 61, 2989 (1974).
147) Fajans, K.: Physical Methods of Organic Chemistry (3. ed.) 1, part II, 1169. New York: Interscience 1960.
148) Salzmann, J. J., Jørgensen, C. K.: Helv. Chim. Acta 51, 1276 (1968).
149) Heydweiller, A.: Physikal. Z. 26, 526 (1925).
150) Jørgensen, C. K.: Theoret. Chim. Acta Chim. Acta 34, 189 (1974).
151) Turner, D. W., Baker, C., Baker, A. D., Brundle, C. R.: Molecular Photoelectron Spectroscopy. London: Wiley-Interscience 1970.
152) Siegbahn, K., Nordling, C., Johansson, G., Hedman, J., Hedén, P. F., Hamrin, K., Gelius, U., Bergmark, T., Werme, L. O., Manne, R., Baer, Y.: ESCA Applied to Free Molecules. Amsterdam: North-Holland Publishing Co. 1969.
153) Jørgensen, C. K.: Chem. Phys. Letters 3, 380 (1969).
154) Bergerhoff, G.: Angew. Chem. 76, 697 (1964).
155) Ruedenberg, K.: Rev. Mod. Phys. 34, 326 (1962).
156) Slater, J. C.: Adv. Quantum Chem. 6, 1 (1972).
157) Johnson, K. H.: Adv. Quantum Chem. 7, 143 (1973).
158) Connolly, J. W. D., Siegbahn, H., Gelius, U., Nordling, C.: J. Chem. Phys. 53, 4265 (1973).
159) Stammreich, H., Kawai, K., Sala, O., Krumholz, P.: J. Chem. Phys. 35, 2168 and 2175 (1961).
160) Cotton, F. A.: Helv. Chim. Acta Fasc. extr. Alfred Werner 117 (1967).
161) Trabelsi, M., Loutellier, A., Bigorgne, M.: J. Organometal. Chem. 40, C 45 (1972).
162) Hayter, R. G., Humiec, F. S.: Inorg. Chem. 2, 306 (1963).
163) Verkade, J. G., Piper, T. S.: Inorg. Chem. 1, 453 (1962) and 2, 944 (1963).
164) Lucken, E. A. C.: Structure and Bonding 6, 1 (1969); Jørgensen, C. K.: Structure and Bonding 6, 94 (1969).
165) Colton, R., Canterford, J. H.: Halides of the First Row Transition-Metals. New York: Wiley-Interscience 1969.
166) Jørgensen, C. K.: Halogen Chemistry (ed.: V. Gutmann). 1, 265. London: Academic Press 1967.
167) Teegarden, K., Baldini, G.: Phys. Rev. 155, 896 (1967).
168) Body, R. G., McClure, D. S., Clementi, E.: J. Chem. Phys. 49, 4916 (1968).
169) Ammeter, J., Schlosnagle, D. C.: J. Chem. Phys. 59, 4784 (1973).
170) Jørgensen, C. K., Berthou, H., Balsenc, L.: J. Fluorine Chem. 1, 327 (1972).
171) Jørgensen, C. K.: Adv. Quantum Chem. 8, 137 (1974).
172) Bjerrum, N., Acta Chem. Scand. 12, 945 (1958).
173) Moore, C. E.: Ionization Potentials and Ionization Limits Derived from the Analyses of Optical Spectra. NSRDS-NBS 34. Washington: National Bureau of Standards 1970.
174) Rosseinsky, D. R.: Chem. Rev. 65, 467 (1965).
175) Latimer, W. M.: J. Chem. Phys. 23, 90 (1955).

176) Schwarzenbach, G.: Pure Appl. Chem. *24*, 307 (1970).
177) Landis, T., Schwarzenbach, G.: Chimia (Aarau) *23*, 146 (1969).
178) Scrocco, E., Salvetti, O.: Ric. Scient. *24*, 1258 and 1478 (1954).
179) Jørgensen, C. K.: Chimia (Aarau) *23*, 292 (1969).
180) Penneman, R. A., Mann, J. B., Jørgensen, C. K.: Chem. Phys. Letters *8*, 321 (1971).
181) Ahrland, S.: Chem. Phys. Letters *2*, 303 (1968).
182) Evans, S., Green, J. C., Jackson, S. E., Higginson, B.: J. Chem. Soc. (Dalton) *1974*, 304.
183) Morris, D. F. C.: Structure and Bonding *4*, 63 (1968) and *6*, 157 (1969).
184) Rabinowitch, E., Thilo, E.: Periodisches System, Geschichte und Theorie. Stuttgart: Ferdinand Enke 1930.
185) Klopman, G.: J. Amer. Chem. Soc. *86*, 1463 and 4550 (1964); ibid. *87*, 3300 (1965).
186) Klopman, G.: J. Amer. Chem. Soc. *90*, 223 (1968).
187) Jørgensen, C. K.: Angew. Chem. *85*, 1 (1973).
188) Wertheim, G. K., Rosencwaig, A., Cohen, R. L., Guggenheim, H. J.: Phys. Rev. Letters *27*, 505 (1971).
189) Jørgensen, C. K., Berthou, H.: Mat. fys. Medd. Danske Vid. Selskab *38*, no. 15 (1972).
190) Jørgensen, C. K., Structure and Bonding *13*, 199 (1973).
191) Jørgensen, C. K., Pappalardo, R., Schmidtke, H. H.: J. Chem. Phys. *39*, 1422 (1963).
192) Jørgensen, C. K.: Chimia (Aarau) *27*, 203 (1973).
193) Schäffer, C. E.: Pure Appl. Chem. *24*, 361 (1970); Structure and Bonding *5*, 68 (1968).
194) Nixon, J. F.: J. Chem. Soc. (Dalton) *1973*, 2226.
195) Campagna, M. Bücher, E., Wertheim, G. K., Buchanan, D. N. E., Longinotti, L. D.: Phys. Rev. Letters *32*, 885 (1974).
196) Ley, L., Pollak, R. A., McFeely, F. R., Kowalczyk, S. P., Shirley, D. A.: Phys. Rev. *B9*, 600 (1974).
197) Jørgensen, C. K.: Chem. Phys. Letters *11*, 387 (1971).
198) Fadley, C. S., Hagström, S. B. M., Klein, M. P., Shirley, D. A.: J. Chem. Phys. *48*, 3779 (1968).
199) Aarons, L. J., Guest, M. F., Hall, M. B., Hillier, I. M.: Trans. Faraday Soc. (II) *69*, 563 (1973).
200) Jørgensen, C. K.: Chimia (Aarau) *25*, 213 (1971).
201) Jørgensen, C. K., Berthou, H.: Chem. Phys. Letters *31*, 416 (1975).
202) Jørgensen, C. K., Berthou, H.: Discuss. Faraday Soc. *54*, 269 (1973).
203) Berthou, H., Jørgensen, C. K.: Analyt. Chem. *47*, 482 (1975).
204) Jørgensen, C. K., Berthou, H.: J. Fluorine Chem. *2*, 425 (1973).
205) Citrin, P. H., Hamann, D. R.: Chem. Phys. Letters *22*, 301 (1973); Phys. Rev. *B 10*, 4948 (1974).
206) Manne, R., Åberg, T.: Chem. Phys. Letters *7*, 282 (1970).
207) Jørgensen, C. K., Judd, B. R.: Mol. Phys. *8*, 281 (1964).
208) Henrie, D. E., Choppin, G. R.: J. Chem. Phys. *49*, 477 (1968).
209) Gruen, D. M.: Progress Inorg. Chem. *14*, 119 (1971).
210) Peacock, R. D.: Structure and Bonding *22*, 81 (1975).

Received July 29, 1974

Stereochemistry of the Reactions of Optically Active Organometallic Transition Metal Compounds

Prof. Dr. Henri Brunner

Fachbereich Chemie der Universität Regensburg, Regensburg

Contents

1. Introduction

Optically active organometallic compounds in which the transition metal is the center of chirality have been known since 1969, when the first manganese compounds were reported[1]. In the meantime cyclopentadienyl and carbonyl transition metal complexes with 4, 5 and 6 ligands have been obtained in optically active form for the following types of compounds (Scheme 1):

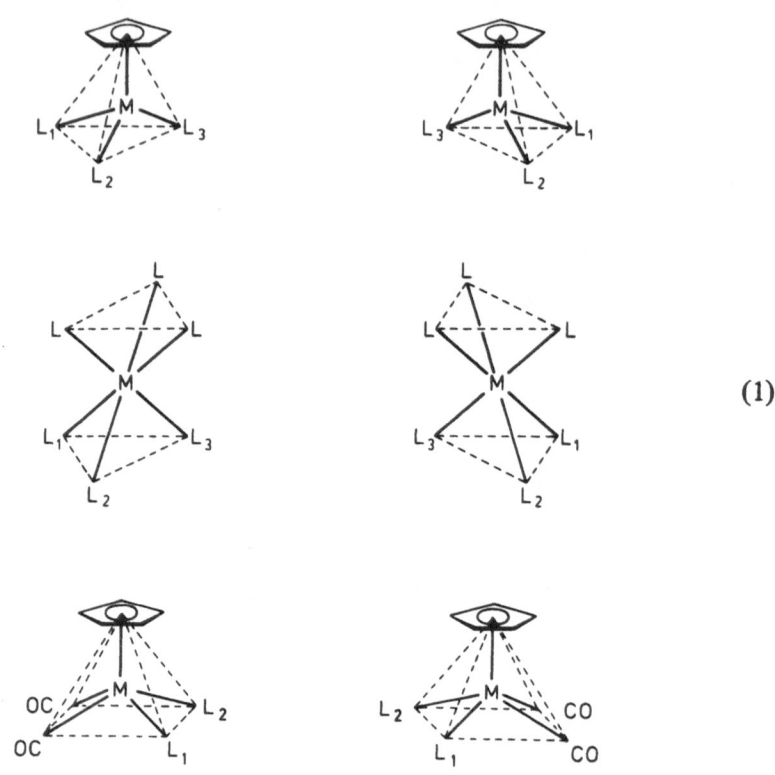

$$(1)$$

The introduction of optically active resolving agents into chiral or prochiral organometallic complexes, the separation of the diastereoisomers and the preparation of enantiomers are summarized in a publication of the New York Academy of Sciences[2]. A brief account of the application of these optically active compounds for the elucidation of reaction mechanisms was given some time ago[3-5]. The present article describes the stereochemical results obtained with the new optically active organometallic compounds, subdivided according to the stereochemical outcome into retention, inversion, racemization and epimerization reactions. The absolute configurations of the optically active organometallic transition metal compounds are not known except for the square pyramidal Mo complex *38a*[6]. Therefore the configurational assignments in the formulas *1–49* are arbitrary, *a* and *b* designating the two different configurations at the metal atom.

2. Retention Reactions

2.1. Ligand Transformations

The optically active compounds obtained by diastereoisomer separations[2] can be used for the preparation of other optically active derivatives. If these reactions occur only at the ligands and do not involve the asymmetric center at the metal atom, all the compounds prepared in this way must have the same relative configurations at the metal atom.

In the reaction of the optically active manganese menthyl esters *1* with sodium methoxide in methanol, a transesterification occurs [Eq. (2)]. The menthoxide ion is replaced by the methoxide ion. In the solvent methanol the equilibrium lies completely on the side of the methyl derivatives *2*, which can be isolated in high yields[7,8].

$$ON \underset{P(C_6H_5)_3}{\overset{Mn}{\longleftarrow}} C \overset{O}{\underset{O}{\diagdown}} + CH_3OH \underset{}{\overset{CH_3ONa}{\rightleftharpoons}} HO\text{---} + ON \underset{P(C_6H_5)_3}{\overset{Mn}{\longleftarrow}} C \overset{O}{\underset{OCH_3}{\diagdown}} \qquad (2)$$

1a $[\alpha]_{579}^{25}$ +485°   *2a* $[\alpha]_{579}^{25}$ +630°

1b $[\alpha]_{579}^{25}$ −550°   *2b* $[\alpha]_{579}^{25}$ −645°

If the (+)-rotating menthyl ester *1a* is used for this reaction the (+)-methyl ester *2a* is obtained. In the same way the (−)-menthyl ester *1b* can be converted to the (−)-methyl ester *2b*[7,8]. Short reaction times and fast work up are necessary because the menthyl esters as well as the methyl and ethyl derivatives racemize in solution[7,8] (Section 4.1).

Treatment of compounds which contain metal-bound ester groups with acids leads to a rapid cleavage of the carbon alkoxide bond in the COOR group[9,10]. If dry HCl is passed through solutions of the optically active esters *1*, the chlorides of the $[C_5H_5Mn(CO) (NO)P(C_6H_5)_3]^+$ cations precipitate within minutes[11]. In this reaction menthol is eliminated and the ester group is converted into the carbonyl group. The chlorides can be transformed into the water-insoluble hexafluorophosphates *3* [Eq. (3)][11].

$$ON \underset{P(C_6H_5)_3}{\overset{Mn}{\longleftarrow}} C \overset{O}{\underset{O}{\diagdown}} + HCl \underset{-Cl^-}{\overset{+ PF_6^-}{\longrightarrow}} \left[ON \underset{P(C_6H_5)_3}{\overset{Mn}{\longleftarrow}} CO \right]^+ PF_6^- + HO\text{---}$$

1a $[\alpha]_{579}^{25}$ +485°   *3a* $[\alpha]_{579}^{25}$ +375°

1b $[\alpha]_{579}^{25}$ −550°   *3b* $[\alpha]_{579}^{25}$ −385°   (3)

If the (+)-rotating ester *1a* is used for this sequence of reactions the (+)-rotating salt *3a* is produced. In the same way, starting with the (−)-rotating ester *1b* the (−)-salt *3b* is obtained[11]. The reactions 3 are best carried out at −20 °C in toluene solution in order to suppress the racemization of the menthyl esters *1*. The salts *3* are optically stable in the solid state as well as in solution. As *3a* and *3b* are also stable to air and heat, they can be stored indefinitely and represent suitable starting materials for the preparation of other optically active compounds, as the following examples show.

The reverse of the acid cleavage of metal-bound ester groups is the nucleophilic addition of alkoxide ions to cationic carbonyl complexes[9]. These reactions have been carried out with the optically active salts *3* and sodium methoxide[7,8] [Eq. (4)] as well as sodium ethoxide[12].

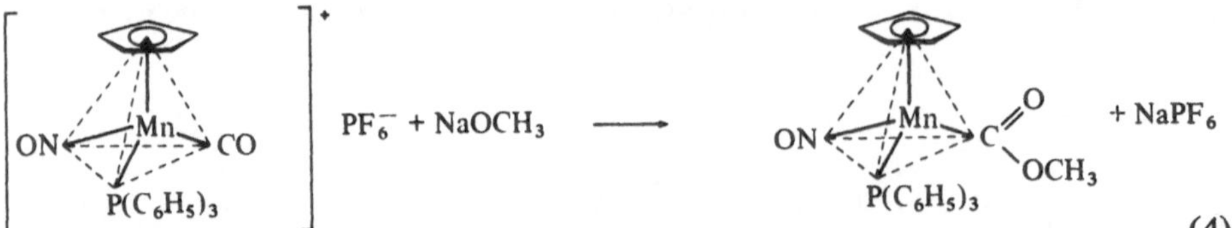

$$3a \ [\alpha]_{579}^{25} +375°$$
$$3b \ [\alpha]_{579}^{25} -385°$$

$$2a \ [\alpha]_{579}^{25} +630°$$
$$2b \ [\alpha]_{579}^{25} -645°$$

(4)

In reaction 4 the (+)-rotating salt *3a* yields the (+)-methyl ester *2a*, identical with the transesterification product of the (+)-menthyl ester *1a* according to Eq. (2). Similarly, the (−)-rotating salt *3b* gives the (−)-rotating methyl ester *2b*, identical with the transesterification product of the (−)-menthyl ester *1b* according to Eq. (2)[7,8].

By reactions 2−4 the (+)-rotating compounds *1a, 2a* and *3a* can be converted into each other. As the reactions occur only at the carbonyl carbon atom group these (+)-rotating compounds must have the same relative configuration at the manganese atom. The (−)-rotating complexes *1b, 2b* and *3b* are related in the same way.

The optically active Fe complex *4a* was converted into the corresponding iodo derivative *5a* with the same configuration at the Fe atom by interaction with HI [Eq. (5)][13].

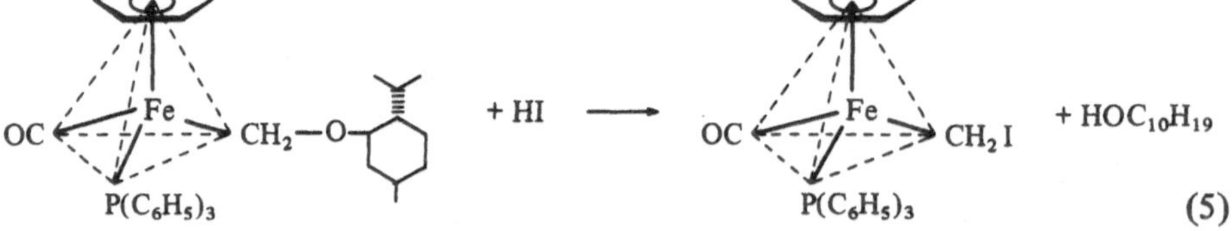

4a *5a*

In the reaction of the salts *3* with Li organic compounds two different types of products are formed simultaneously because LiR attacks cyclopentadienyl carbonyl cations at two different sites: on the cyclopentadienyl ring, and on the carbonyl carbon atom[14−16]. In the ring-addition reaction the cations are converted into the neutral exo-cyclopentadiene complexes 6 whereas in the carbonyl-addition reaction the cations are transformed to neutral acyl complexes 7[14].

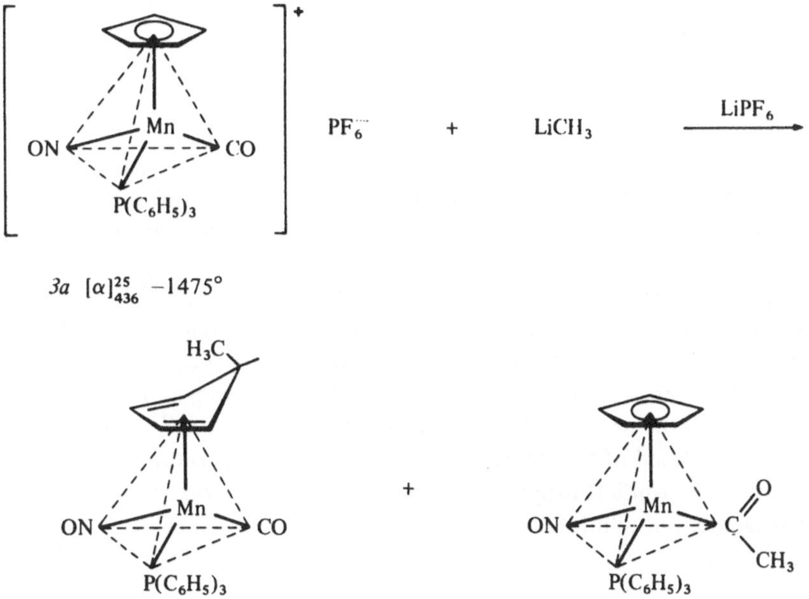

3a $[\alpha]^{25}_{436}$ −1475°

(6)

6a $[\alpha]^{25}_{436}$ −288°

7a $[\alpha]^{25}_{436}$ −2450°

The reaction of 3a with LiCH₃ is shown in Eq. (6)[17]. In the same way the optically active salts 3a and 3b were reacted with Li phenyl and p-substituted Li phenyls with the p-substituents $N(CH_3)_2$, OCH_3, CH_3, C_6H_5, F, Cl, CF_3[17]. The cyclopentadiene complexes 6a and 6b and those carrying other exo substituents are optically stable like the salts 3a and 3b[17]. The acetyl compounds 7a and 7b as well as the corresponding benzoyl and p-benzoyl derivatives 18−25, however, are configurationally labile in solution[17] (Section 4.1).

In reaction 6 and in the reactions of 3a and 3b with other Li organic compounds, the two products must necessarily have the same relative configuration at the metal atom as the optically active cations used for their preparation, because the asymmetric center at the metal atom does not take part in these ligand transformations. This argument holds for reactions 2−6, but it is not always reliable, as the examples in Section 3.1 show. It should be mentioned that compounds with the same relative configuration at the manganese atom have similar chiroptical properties. So the wavelengths of the maxima and the signs of the Δε values in the CD spectra of benzoyl compounds with different substituents are exactly the same and they are not too different in compounds in which one ligand is varied a little[17]. Therefore it seems that the chiroptical properties of these organometallic compounds are dominated by the configura-

71

tion at the metal atom and that small changes in the ligands do not have a large influence on ORD and CD spectra.

2.2. Ligand Dissociation via Chiral Intermediates

Another group of retention reactions including ligand dissociation and the formation of chiral intermediates is discussed in Section 4.1.

2.3. SO₂ Insertion

SO_2 inserts into metal-carbon σ bonds[18]. The reaction is stereospecific with respect to the configuration at the α-carbon atom of the alkyl group[19] and proceeds with inversion of configuration at the α-carbon atom, as shown on *threo-erythro* isomeric derivatives of $C_5H_5Fe(CO)_2CHDCHDC(CH_3)_3$ by ^1H NMR spectroscopy[20,70].

As far as the metal center is concerned, the stereospecificity of the SO_2 insertion was demonstrated with diastereoisomeric Fe compounds that contain chiral centers on the metal and either in the substituted cyclopentadienyl ring[21] or in the alkyl side-chain[21, 22]. The stereochemistry of the SO_2 insertion with respect to the metal atom was studied with the optically active iron compounds *8a* and *8b*[23]. Liquid SO_2 inserts into the Fe–C bond of *8a* and *8b* with retention of configuration[23] [Eq. (7)]. The similarity in the chiroptical properties of *8a* and *9a* as well as *8b* and *9b* was used for the assignment of relative configurations[23].

8a $[\alpha]_{578}^{25}$ +330°

8b $[\alpha]_{578}^{25}$ −309°

9a $[\alpha]_{578}^{25}$ +206°

9b $[\alpha]_{578}^{25}$ −270°

(7)

These results lead to the following suggestion for the mechanism of the SO_2 insertion into a metal-alkyl bond[18]: Backside attack of SO_2 on the α-carbon atom of the alkyl group with inversion of configuration at the carbon atom affords a tight ion pair $M^+RSO_2^-$, from which first the O- and then the S-bonded products MSO_2R are formed with retention of configuration at the metal atom[18, 23]. The CO insertion reaction is discussed in Section 3.3.

3. Inversion Reactions

3.1. Role Change of Ligands

In the transesterification of the manganese menthyl esters *1a* and *1b*, which give the corresponding methyl derivatives *2a* and *2b*, the same relative configuration was ascribed to starting materials and reaction products, the argument being that the reactions occur only at the ligand, the configuration at the metal atom being unchanged[7, 8]. The iron menthyl esters *10a* and *10b*[24, 25] differ from the manganese menthyl esters *1a* and *1b* only in the fact that the manganese nitrosyl system is replaced by the isoelectronic iron carbonyl system[26], everything else remaining unchanged. Therefore, in the transesterification reaction of the iron menthyl compounds *10*, the corresponding methyl esters *11* with the same configuration at the iron atom were to be expected. The transesterification in methanol according to Eq. (8) in fact yields the methyl ester, but always in the racemized form *11a/b*[25]. Furthermore, for the transesterification in the manganese series base addition is necessary, whereas the transesterification in the iron series proceeds in pure methanol[8, 25].

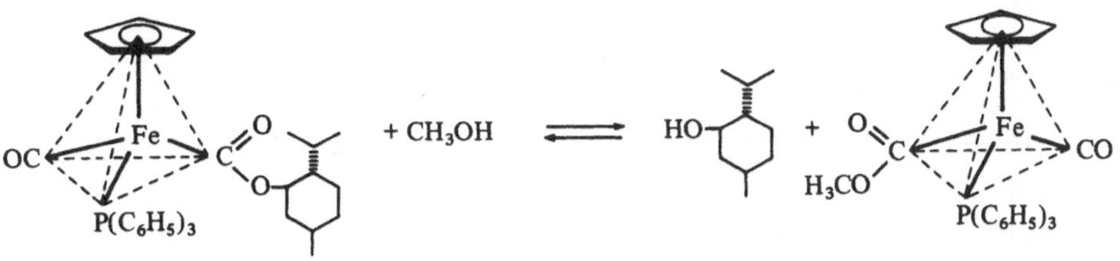

$10a \ [\alpha]_{579}^{25} \ +35°$

$\underline{10b} \ [\alpha]_{579}^{25} \ -80°$

11 racemic

$$(8)$$

These differences in the reaction conditions and the stereochemical results of the transesterification of the related manganese and iron systems of Eqs. (2) and (8) are explained on the basis of the results of the following reaction.

In the reaction of the optically active iron menthyl esters *10* with Li methyl, the esters are converted into the acetyl derivatives *12* and Li menthoxide is eliminated[27]. If in reaction 9 the (+)-rotating ester *10a* is used, the (−)-rotating acetyl derivative *12b* is formed and the (−)-rotating ester *10b* affords the (+)-rotating acetyl derivative *12a*[27].

Contrary to the examples in Section 2.1, the optical rotations of starting material and reaction product in reaction 9 are not the same·and their chiroptical properties are also opposite to each other. The diagram at the bottom of Scheme 9 shows the CD spectra of *10a* and its reaction product with Li methyl *12b*[27]. These opposite Cotton effects indicate that the configuration at the iron atom is inverted in reaction 9. This can be explained in the following way: Li methyl obviously does not, as expected, attack the carbon atom of the ester group, but the carbon atom of the carbonyl group. However, the menthoxide ion is eliminated from the ester group, which leads to an inversion of the configuration at the iron atom because the original carbonyl group is transformed into the new acetyl group, whereas the original functional group on losing the menthoxide ion is converted into the new carbonyl group[27]. In this reaction two ligands change their roles and this means inversion of configuration, although the bonds from the iron atom to the ligands are not cleaved. Obviously the carbonyl group in compounds *10* is more reactive than the ester group towards nucleophilic attack[25, 27].

10a $[\alpha]^{25}_{546}$ +72°

10b $[\alpha]^{25}_{546}$ −120°

12b $[\alpha]^{25}_{546}$ −289°

12a $[\alpha]^{25}_{546}$ +287°

$$\tag{9}$$

In the manganese compounds *1* the ester group seems to be more reactive than the nitrosyl group and the nucleophilic attack proceeds at the carbon atom with retention of configuration[7, 8]. Therefore, for molecules of type *1* and *10* the reactivity sequence toward nucleophilic attack is[25, 27]:

$$CO > COOR > NO. \tag{10}$$

This reactivity sequence also explains the results of the transesterification reactions 2 and 8. Nucleophilic attack on the iron complex *10* is easier at the carbonyl group than at the ester group. One reaction step of this kind leads to an inversion of configuration, as shown in Eq. (9)[27]. Successive nucleophilic attacks, as occur in the solvent methanol, however, lead to a series of inversion steps and ultimately to the racemization observed in reaction 8[25]. In the manganese ester *1*, on the other hand, the attack of the alkoxide ion occurs at the carbon atom of the more reactive ester group and hence always with retention of configuration[8]. In accord with reactivity sequence 10, pure methanol is sufficient to attack the reactive CO group in the transesterification of *10* [Eq. (8)] whereas Na alkoxide must be added to attack the less reactive COOR group in the transesterification of 1 [Eq. (2)][8, 25].

3.2. Walden Inversion

The inversion reactions 8 and 9 of the iron esters *10* discussed in Section 3.1 are inversions of a special kind. They arise because two ligands change roles without cleavage of the bonds starting from the metal atom. The next examples show an inversion reaction that occurs directly at the metal atom.

The optically active Mo complexes *13a* and *13b* are configurationally stable[28, 29]. Their optical rotations in solution remain constant over long periods of time. If, however, a trace of free R—(—)—α-phenyl ethyl isonitrile[30, 31] is added to solutions of *13a* at room temperature, its optical rotation decreases within about one hour to values around O[28, 29]. We explain this behavior by a backside attack of the free isonitrile on the optically active complex *13a* according to Eq. (11).

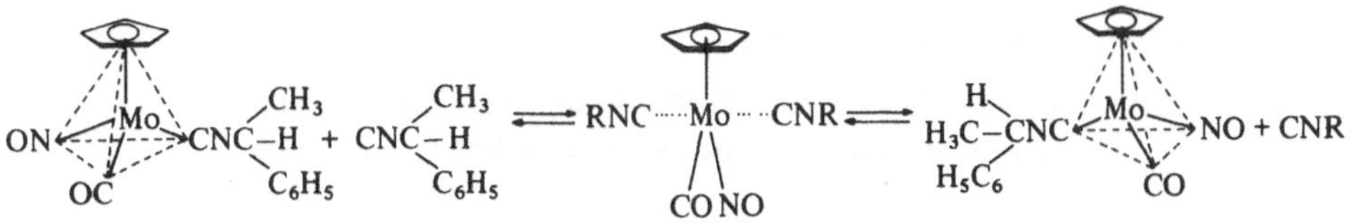

13a $[\alpha]_{546}^{25}$ +350°

13b

R = CH (CH₃) (C₆H₅) (11)

The isonitrile originally bound in the complex is expelled on the other side and a Walden inversion occurs at the asymmetric Mo atom[28, 29]. Complex *13b*, with the opposite configuration to *13a* at the Mo atom, may react again with free isonitrile via a similar S_N2 transition state. A series of such inversion steps ultimately leads to the epimerization which results in the observed decrease in optical rotation on addition of free α-phenyl ethyl isonitrile to solution of *13a* or *13b* [28, 29].

The optically active Mo isonitrile complex *13a* may also be reacted with nucleophiles other than isonitriles. Triethylphosphine, a much better nucleophile than the isonitrile bound in *13a*, also attacks complex *13a* at the backside and replaces the isonitrile via an S_N2 transition state in a reaction which at room temperature is completed within a few minutes[28, 29].

The triethylphosphine derivative *14b* formed in reaction 12 has a configuration opposite to that of the starting material, as can be seen by comparing the chiroptical properties of *13a* and *14b* and of the products of Scheme 12 when cyclohexyl isonitrile is used instead of triethylphosphine[28, 29].

System 12 is very favorable, since the optically active phosphine complex *14b* is relatively stable towards the free nucleophiles, the excess phosphine, on the one hand and the isonitrile formed in reaction 12 on the other. Therefore, if the reaction of *13a* with $P(C_2H_5)_3$ is stopped after a few minutes, only the inversion represented

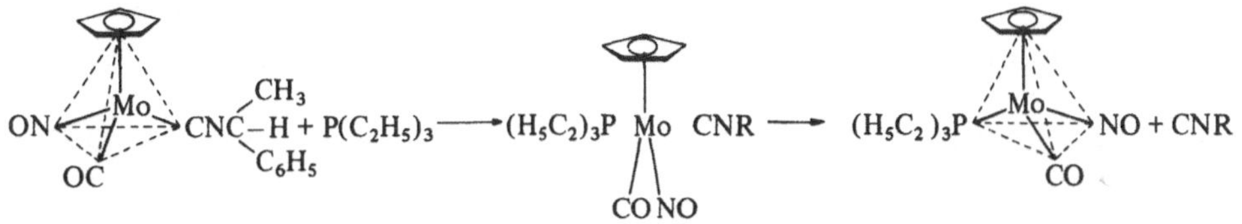

13a $[\alpha]_{546}^{25}$ +350°

14b $[\alpha]_{546}^{25}$ −215°

R = CH (CH₃) (C₆H₅) (12)

75

in Eq. (12) will have occurred and the phosphine complex *14b* can be isolated in an optically active form[28, 29].

The reactions 11 and 12 of the optically active Mo compounds *13* follow the typical pattern of organic S_N2 reactions. Racemization occurs if the optically active compounds are treated with nucleophiles that are identical with the leaving groups. Better nucleophiles lead to a Walden inversion.

3.3. CO Insertion

Metal-alkyl bonds can be carbonylated and metal-acyl bonds can be decarbonylated[32]. The main mechanism suggested for the carbonylation reaction is alkyl migration from the metal to a terminal CO group with entrance of a ligand L, and vice versa for decarbonylation[32].

Both reactions have been shown to proceed stereospecifically with respect to the carbon atom adjacent to the metal or the carbonyl group[19, 33]. An investigation of the *threo-erythro* isomers of $C_5H_5Fe(CO)_2CHDCHDC(CH_3)_3$ showed that the reaction occurs with retention of configuration at the α-carbon atom of the alkyl chain[34]. The same conclusion was drawn from a reaction sequence involving carbonylation of an Fe alkyl compound with the asymmetric center on the α-carbon atom followed by an oxidative cleavage of the Fe-acyl bond with halogen[35].

The photochemical decarbonylation of $1.3-(CH_3)(C_6H_5)C_5H_3Fe(CO)-[P(C_6H_5)_3]COCH_3$ proceeds with a high degree of stereospecificity as far as the configuration at the asymmetric iron center is concerned[22, 36, 71]. The loss in stereospecificity was shown to be due to epimerization subsequent to CO elimination. Similar results have been obtained with diastereoisomers of $C_5H_5Fe(CO)[P(C_6H_5)_3]-CH_2CH(CH_3)(C_6H_5)$[22, 36, 71].

The thermal decarbonylation of $(-)-C_5H_5Fe(CO)[P(C_6H_5)_3]COCH_3$ *12b*[37] in boiling toluene gave only optically inactive $C_5H_5Fe(CO)[P(C_6H_5)_3]CH_3$ *15*[38]. The photochemical decarbonylation of $(+)-C_5H_5Fe(CO)[P(C_6H_5)_3]COCH_3$ *12a*[39] gave on long irradiation an optically inactive decarbonylation product, and on short irradiation an optically active decarbonylation product *15b* to which the opposite configuration at the iron atom was assigned on the basis of its chiroptical properties[38] [Eq. (13)].

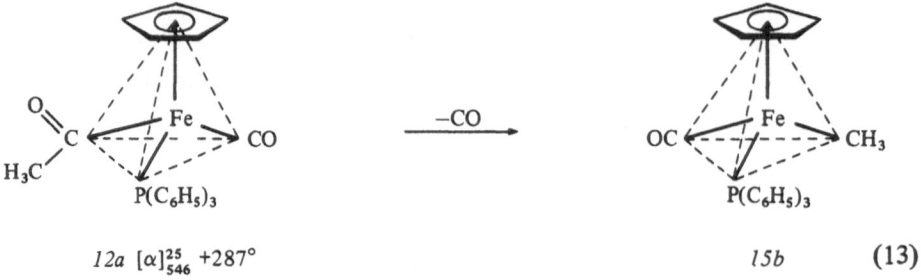

$12a$ $[\alpha]_{546}^{25}$ $+287°$ $15b$ (13)

The inversion of configuration at the iron atom in Eq. (13) is in accord with the proposed alkyl migration mechanism for the decarbonylation reaction[22, 32, 72].

4. Racemization and Epimerization Reactions

All the new optically active organometallic complexes[2] are configurationally stable in the solid state. The crystals can be stored indefinitely without any decline in the optical rotations, provided no decomposition reactions occur. In solution, however, two different cases must be distinguished. Most of the new organometallic compounds are also optically stable in solution, and the solutions retain their optical rotations unchanged for long periods of time. But some complexes are configurationally labile in solution; their rotational values decrease without participation of other reagents. Examples of this kind are treated in Sections 4.1 and 4.2. Section 4.3 describes the racemization and epimerization reactions that occur when optically active organometallic complexes interact with other reagents.

4.1. Dissociation Reactions

After dissolution of the manganese methyl esters in benzene at 30 °C specific rotations of about 2500° can be measured for the (+)- and (−)-rotating enantiomers $2a$ and $2b$ [7, 8]. These rotations decrease exponentially with time and O is approached from both sides (Fig. 1). The decrease in rotation is a first-order reaction [7, 8]. The half-life of the optical stability of the methyl esters $2a$ and $2b$ at 30 °C in benzene solution is 2 h 50 min. During the racemization reaction the configuration at the manganese atom changes and an equilibrium is established between the two enantiomers $2a$ and $2b$. On completion of racemization, equal amounts of enantiomers $2a$ and $2b$ are present [7, 8]. In contrast to the manganese esters $C_5H_5Mn(COOR)(NO)$-$P(C_6H_5)_3$, the isoelectronic iron esters $C_5H_5Fe(COOR)(CO)P(C_6H_5)_3$ do not racemize in solution [24, 25].

The strongly positive entropy of activation for the racemization of $2a$ and $2b$ (25 eu). indicates that a dissociative mechanism is operative [7]. As triphenylphosphine is a stable molecule and triphenylphosphine dissociations are known in other systems, the cleavage of the Mn−P bond could be responsible for the racemization. Therefore we studied the influence of triphenylphosphine on the racemization of $2a$ and $2b$ and the phosphine exchange in 2.

The decline in optical rotation of the esters $2a$ and $2b$ as a function of time is not affected by the addition of triphenylphosphine (Fig. 1). $P(C_6H_5)_3$ in concentrations up to 10-fold excess with respect to complex concentration had no effect at all on the rate of racemization of $2a$ and $2b$ [7]. Deuteration experiments indicated that triphenylphosphine is indeed the ligand which dissociates. When the racemization of the manganese esters 2 is carried out in the presence of deuterated triphenylphosphine, $P(C_6D_5)_3$ is incorporated into the complexes 16 [Eq. (14)]. A kinetic study showed that the rate of triphenylphosphine exchange is exactly equal to the rate of racemization [8].

All these results are consistent with the S_N1 mechanism shown in Eq. (15). The Mn−P bond in $2a$ is cleaved in the rate-determining step k_1 and a planar intermediate 17 is formed.

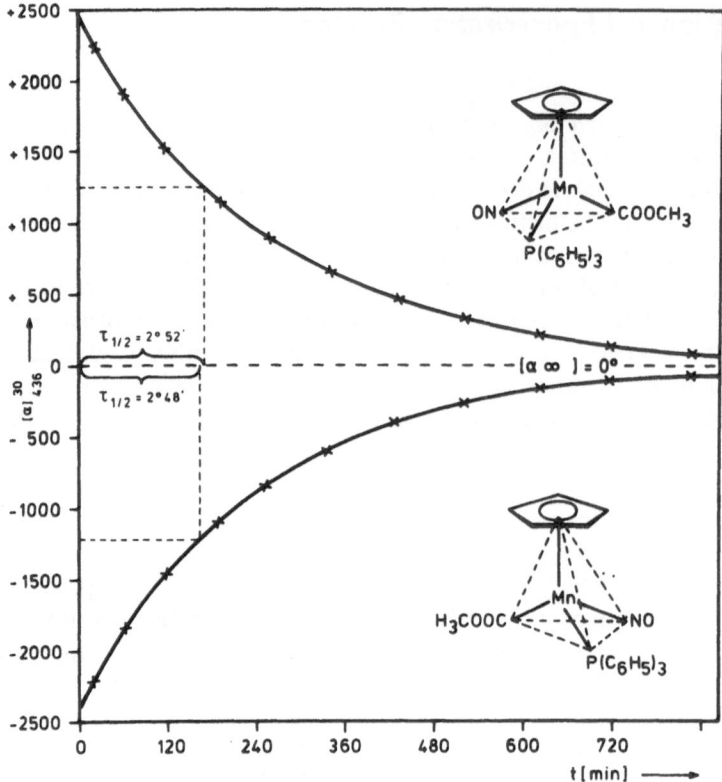

Fig. 1. Racemization of (+)- and (−)-C$_5$H$_5$Mn(COOCH$_3$)(NO)P(C$_6$H$_5$)$_3$ *2a* and *2b* in benzene solution at 30 °C in the absence and in the presence of P(C$_6$H$_5$)$_3$

17 is attacked by triphenylphosphine in a rapid reaction step k_2. As this attack may occur with equal probability at the front or back side of *17*, the enantiomers *2a* and *2b* are formed in equimolar amounts; phosphine exchange is also possible at this stage. As triphenylphosphine does not take part in the rate-determining step k_1, the rate of racemization of *2a* and *2b* is not affected by the presence of triphenylphosphine[7, 8] (Fig. 1). The simple dissociation mechanism of Eq. (15) has, however, to be modified after consideration of the manganese benzoyl compound *21a*[17, 40].

The rates and half-lives for the racemization of the C$_5$H$_5$Mn(COOR)(NO)P(C$_6$H$_5$)$_3$ compounds are dependent on the group R in the ester function. As the alkyl group increases from methyl to ethyl to menthyl, racemization becomes faster and the

$$\underset{2}{\text{ON}\text{-Mn-COOCH}_3\,\,P(C_6H_5)_3} \quad \underset{+P(C_6H_5)_3\,;\,\,-P(C_6D_5)_3}{\overset{+P(C_6D_5)_3\,;\,\,-P(C_6H_5)_3}{\rightleftarrows}} \quad \underset{16}{\text{ON}\text{-Mn-COOCH}_3\,\,P(C_6D_5)_3} \qquad (14)$$

half-lives at 30 °C in benzene solution fall from 170 to 115 to 60 min. This is certainly due to the increasing size of the alkyl group, which should favor the dissociation of triphenylphosphine. However, this steric effect is superimposed upon an electronic effect[7, 8, 12, 17].

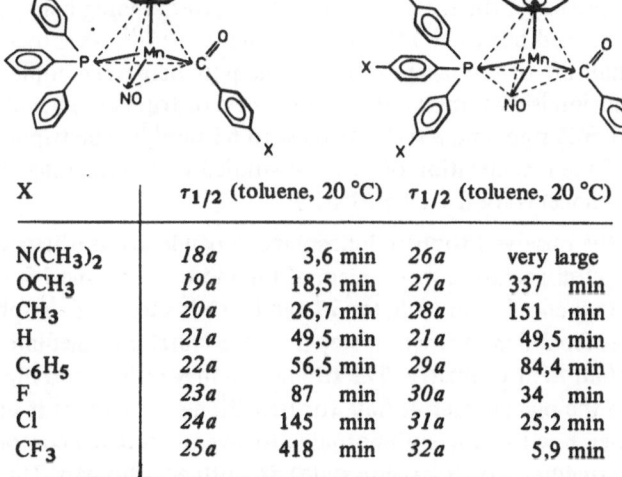

$$2a \qquad\qquad 17 \qquad\qquad 2b$$

$$(15)$$

In order to study the electronic effect without changing the steric situation around the reaction center, we prepared several benzoyl compounds $18-25$ with different p-substituents $X^{14)}$ [Section 2.1, Eq. (6)]. The optically active complexes $18-25a$ and b show in solution the same exponential decline in optical rotation as do the corresponding esters $2a$ and $2b$. The left column of Fig. 2 gives the half-lives for these first-order reactions of the benzoyl derivatives $18a-25a$, the half-life of the unsubstituted benzoyl compound $21a$ in toluene at 20 °C being 49.5 minutes[17]. Electron-releasing substituents in the p position increase the rate of racemization. The dimethylamino compound $18a$ with $\tau_{1/2}$ 3.6 minutes at 20 °C can be handled without racemization only at low temperatures[17]. Electron-withdrawing substituents in the p position of the benzoyl group decrease the rate of racemization. The racemization of the trifluoromethyl compound $25a$ is fast only at higher temperatures[17].

X		$\tau_{1/2}$ (toluene, 20 °C)		$\tau_{1/2}$ (toluene, 20 °C)
$N(CH_3)_2$	$18a$	3,6 min	$26a$	very large
OCH_3	$19a$	18,5 min	$27a$	337 min
CH_3	$20a$	26,7 min	$28a$	151 min
H	$21a$	49,5 min	$21a$	49,5 min
C_6H_5	$22a$	56,5 min	$29a$	84,4 min
F	$23a$	87 min	$30a$	34 min
Cl	$24a$	145 min	$31a$	25,2 min
CF_3	$25a$	418 min	$32a$	5,9 min

Fig. 2. Effect of p-substituents X in the complexes $C_5H_5Mn(CO-p-C_6H_4X)(NO)P(p-C_6H_4X)_3$ $18a-32a$ on the half-lives $\tau_{1/2}$ of the first-order racemization reaction in toluene at 20 °C (complex concentration 2 mg/ml)

In the same way as the electron density of the manganese atom was varied by use of different p-substituents in the benzoyl group, it should be possible to modify the electron density of the phosphorus atom by introducing substituents X into the three p positions of the triphenylphosphine ligand. The compounds *26a–32a* were prepared in the same way as the corresponding triphenylphosphine complexes[1, 7, 8], the diastereoisomer separations being made via the menthyl esters[41]. From the half-lives of the complexes *26a–32a* in the right column of Fig. 2, it can be seen that p-substitution in the phosphine ligand has exactly the opposite effect from p-substitution in the benzoyl system[41]. It can be concluded that cleavage of the manganese-phosphorus bond is fast when electron-donating substituents are attached to the metal atom and electron-attracting substituents are bound on the phosphorus atom whereas, if the substituent effects are inverted, cleavage of the metal-phosphorus bond is slow[17, 41]. In this context it should be mentioned that the iron-acyl complexes $C_5H_5Fe(COR)(CO)P(C_6H_5)_3$ isoelectronic with the manganese compounds $C_5H_5Mn(COR)(NO)P(C_6H_5)_3$ do not racemize in solution[27]. Moreover, the complex cation in the salt $[C_5H_5Mn(CO)(NO)P(C_6H_5)_3][PF_6]$ is configurationally stable in solution even at higher temperatures[11].

For both series of compounds, *18a–25a* and *26a–32a*, with the exception of the tris-diphenylphosphine compound *29a*, we established good Hammett correlations between the substituent constants and the half-lives of the configurational stability[17, 41]. As Fig. 2 shows, small changes in the ligands cause large differences in the rate of racemization. Therefore the half-lives of the racemization reaction offer a much more sensitive tool for studying changes in the electron density at the manganese or phosphorus atom than do shifts of the IR stretching vibrations or the NMR signals, which are rather small in the compounds *18–32*[17, 41].

The racemization of the optically active manganese methyl esters *2a* and *2b* proceeds with the same rate irrespective of whether triphenylphosphine is present or not[7, 8] (Fig. 1). Surprisingly, the racemization of the corresponding benzoyl complexes *21a* and *21b* turned out to be $P(C_6H_5)_3$-dependent[17, 40]. As shown graphically in Fig. 3, the half-life of the racemization of the pure benzoyl complex *21a* at 20 °C in toluene solution is 49.5 min; a threefold excess of triphenylphosphine prolongs the half-life to 56.5 min, and a sixfold excess to 65 min[17]. The triphenylphosphine dependence of the racemization of *21a* was studied within the range complex: triphenylphosphine concentration 1 : 1 to 1 : 20[17, 40].

To account for the observed triphenylphosphine dependence, it is necessary to expand the simple S_N1 dissociation mechanism of Eq. (15). In Scheme 16, upper half, the introduction of two chiral intermediates *33a* and *33b* is suggested. Triphenylphosphine dissociates from *21a* in reaction step a. In the chiral intermediate *33a* the remaining ligands retain their geometry. The tripod *33a* may either rearrange according to step c to give the planar intermediate (or transition state) *34*, or it may react in the bimolecular step b with triphenylphosphine to give the original compound *21a*. From the planar intermediate (or transition state) *34* both enantiomers *21a* and *21b* can be obtained with equal probability. The fact that reaction step b, the uptake of triphenylphosphine, is favored in the presence of increasing amounts of $P(C_6H_5)_3$ explains the decrease in the racemization rate on addition of $P(C_6H_5)_3$ [17,40].

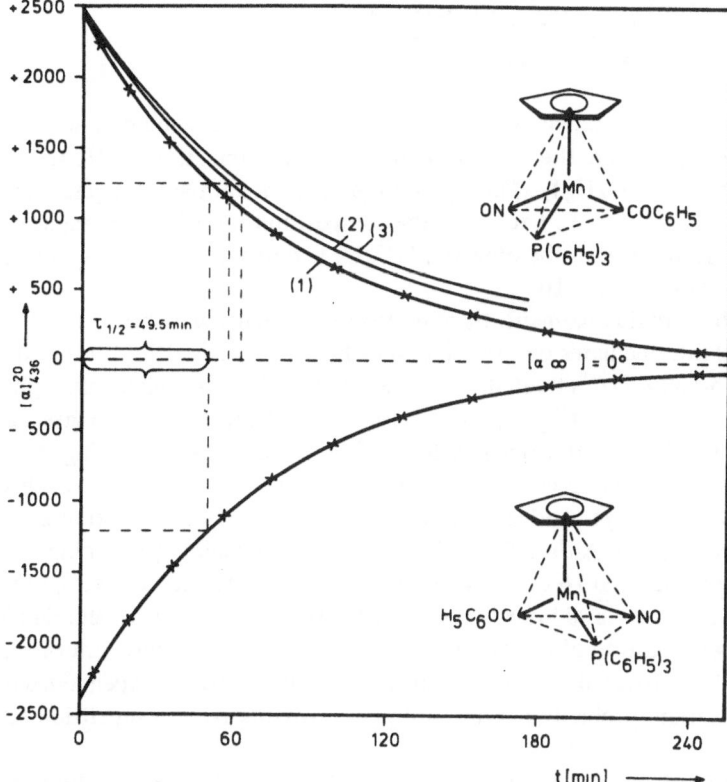

Fig. 3. Racemization of (+)- and (−)-$C_5H_5Mn(COC_6H_5)$ (NO)$P(C_6H_5)_3$ *21a* and *21b* in toluene solution at 20 °C (complex concentration 2 mg/ml).
(1) in absence of $P(C_6H_5)_3$.
(2) in presence of $P(C_6H_5)_3$ in 3-fold excess.
(3) in presence of $P(C_6H_5)_3$ in 6-fold excess.

If the kinetic equation for the upper part of Scheme 16 is set up and the Bodenstein steady-state approximation applied, we have[17,40)]

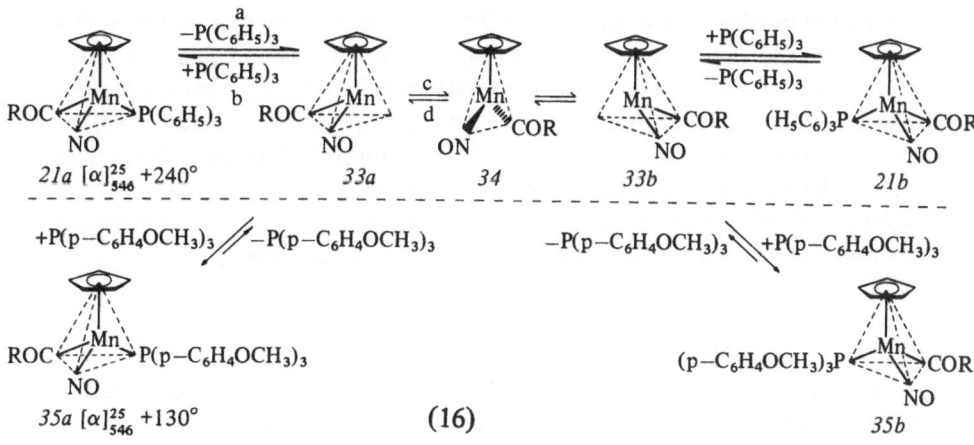

$$(16)$$

$$\frac{k_c}{k_b} = \frac{k_{with} \cdot [P(C_6H_5)_3]}{k_{without} - k_{with}} \tag{17}$$

where k_b and k_c are the rate constants for the reaction steps b and c in Scheme 16, and k_{with} and $k_{without}$ are the rate constants for the racemization of $21a$ with and without addition of $P(C_6H_5)_3$. Throughout the range of complex : triphenylphosphine concentration from 1 : 1 to 1 : 20 the competition ratio k_c/k_b remains constant[17,40]. This means that the observed $P(C_6H_5)_3$ dependence is also quantitatively consistent with Scheme 16.

Besides this kinetic argument, there is also stereochemical evidence for the presence of chiral intermediates on dissociation of $P(C_6H_5)_3$ from $21a$. If a solution of the benzoyl complex $21a$ is treated with an excess of tri-p-anisylphosphine, phosphine exchange takes place and $P(C_6H_5)_3$ is replaced by $P(p-C_6H_4OCH_3)_3$ to give the substitution product 35[40]. After one half-life of $21a$, the unchanged starting material $21a$ and the reaction product 35 can be separated by chromatography[40]. The substitution product 35 (Scheme 16) is optically active and has the same configuration as the $P(C_6H_5)_3$ complex $21a$. Although this is a dissociation reaction, phosphine substitution according to Scheme 16 occurs with at least partial retention of configuration at the Mn atom, because chiral intermediates are formed. Mechanism 16 should also be valid for the racemization of the esters $2a$ and $2b$, but here the chiral intermediates could not be demonstrated on the basis of the triphenylphospine dependence of the rate of racemization, probably because of the unfavorably large competition ratio k_c/k_b [40].

The stability of the chiral intermediates may be ascribed either to π-bonding effects or to solvent participation. Flattening of the chiral intermediate would require at least one ligand to move relative to the metal, a process which could be accompanied by some loss of π-bonding. This may be the reason why the other ligands retain their geometrical arrangement on dissociation of triphenylphosphine, and why it seems reasonable to assume an energy barrier between the chiral intermediate $33a$ and the planar form 34. On the other hand, solvent molecules might be expected to stabilize the free coordination position formed on dissociation of $P(C_6H_5)_3$ since even unreactive molecules are known to interact with free coordination sites[42, 43, 73]. Although the results presented demonstrate the existence of the chiral intermediates $33a$ and $33b$, they do not tell us whether the planar form 34 is an intermediate or a transition state[40].

The kinetic and stereochemical results discussed in this section could be explained in another way by assuming that group R migrates rapidly from the acyl substituent to the metal atom to give an intermediate of the type $C_5H_5Mn(CO)(NO)R$. This intermediate could be formed in an additional equilibrium starting from $33a$ in Scheme 16, or by a concerted process directly from $21a$. Intermediates of the type $C_5H_5Mn(CO)(NO)R$ would also account for the observed retention stereochemistry. Further work is necessary to distinguish between these possibilities.

4.2. Intramolecular Epimerization Reactions

The compounds $C_5H_5M(CO)_3X$ (M = Mo, W; X = alkyl, Hal and so on) have square pyramidal structures with the π-cyclopentadienyl ring on top of the pyramid[44—46].

If a carbonyl group is replaced by a ligand L, *cis-trans* isomers *36* and *37* are formed [46–48]. The isomer interconversion has been studied with the help of NMR spectroscopy [46,49,50] and the intramolecular character of the rearrangement shown in Eq. (18) has been demonstrated [46,51,52]. The rate of isomerization is strongly dependent on the nature of L and X [46,47].

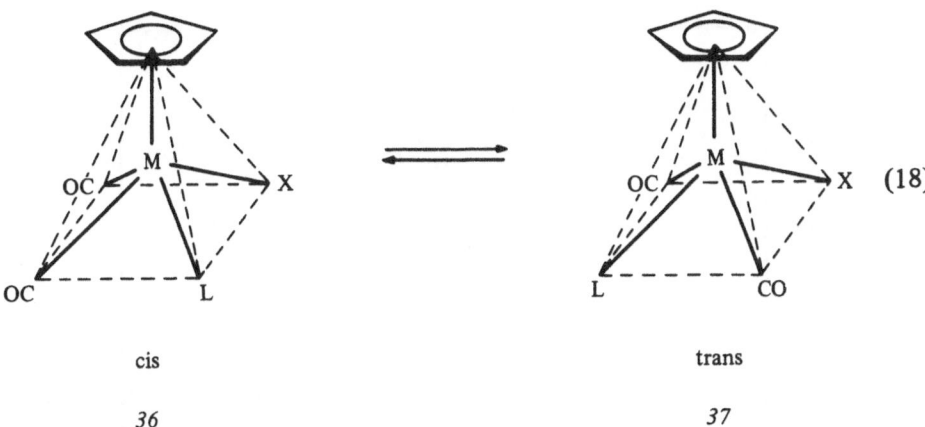

cis   trans

36   *37*

Whereas the *trans* isomers *37* contain a plane of symmetry, the *cis* isomers *36* are asymmetric [46,53]. If L and X are replaced by a chelate ligand LL' with two different ends, *cis-trans* isomerization according to Eq. (18) is excluded and only the enantiomeric *cis* isomers *36* are possible [54].

With the optically active Schiff base derived from pyridine carbaldehyde-(2) and S-(−)-α-phenyl ethyl amine replacing L and X, the two diastereoisomeric complexes, *38a* and *b* for Mo and *39a* and *b* for W, respectively, are formed, differing only in the configuration at the metal atom [54,55]. In both cases the two epimers can be separated by fractional crystallization [55]. For the Mo compound the absolute configuration has been determined, formula *38a* representing the $(+)_{579}$-rotating isomer [6].

At room temperature the optical rotations of the salts *38a* and *b* dissolved in acetone or DMF remain constant, at higher temperatures, however, an exponential decrease of rotation is observed, as shown in Fig. 4 for 75 °C [54–56]. Because the complexes *38a* and *38b* are epimers, the plots do not begin at the same (+)- and (−)-rota-

(19)

M = Mo	*38a*	$[\alpha]^{25}_{579}$ + 905°	M = Mo	*38b*	$[\alpha]^{25}_{579}$ − 820°
W	*39a*	$[\alpha]^{25}_{579}$ + 315°	W	*39b*	$[\alpha]^{25}_{579}$ − 320°

tions, although the compounds *38a* and *b* are optically pure and the curves do not approach 0 but a value different from 0. Nevertheless the decrease of optical rotation for the diastereoisomers *38a* and *b* in the temperature range 55–95 °C is a clean first-order reaction[56]. During the epimerization the configuration at the Mo atom changes and an equilibrium is established between species *38a* and *b*[56] [Eq. (19)].

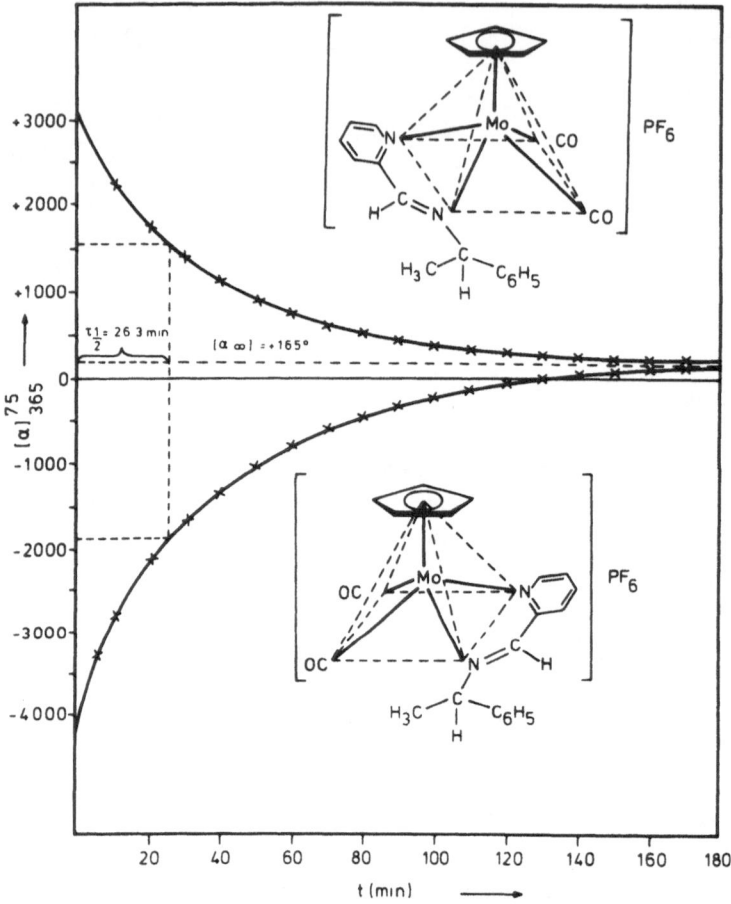

Fig. 4. Epimerization of the Mo complexes *38a* and *b* at 75 °C in dimethylformamide

There are several possibilities to account for the transformation of the diastereoisomers *38a* and *b* into each other. For a compound with 5 substituents an intramolecular mechanism without cleavage of bonds is very attractive, but dissociative mechanisms must also be considered, for instance chelate ring opening and closing to the other side; or ligand dissociation, for instance dissociation of the Schiff base[56]. If the Schiff base of pyridine carbaldehyde-(2) and methyl amine is added during epimerization of *38a*, no exchange is observed with the Schiff base bound in complex *38a*[56]. Therefore Schiff-base exchange can be excluded as a mechanism for the observed epimerization reaction. Furthermore, if the epimerization is carried out in the presence of $P(C_6H_5)_3$, triphenylphosphine does not show up in the reaction product. The inference is that during the change of the configuration at the Mo atom no free coordination positions are available[56]. This evidence favors an intramolecular character for the epimerization, for which several modes can be envisaged[57–59].

In addition to its use in the preparation of the square pyramidal Mo and W complexes *38* and *39*, the Schiff base derived from pyridine carbaldehyde-(2) and S-(−)-α-phenyl ethyl amine[54] was also used for the synthesis of optically active Co complexes of the tetrahedral type[60]. Unlike the Mo and W compounds, the separated Co diastereoisomers *40a, 40b* one of which is shown in Scheme 20, are optically stable. The rigidity of the tetrahedral Co complexes and the nonrigidity of the square pyramidal Mo and W complexes give a further indication of the intramolecular character of the epimerization of *38* and *39*.

$$40a \; [\alpha]_{579}^{25} + 150° \qquad 38a \; [\alpha]_{579}^{25} + 905° \tag{20}$$

Analogous behavior was found as regards the isomerization of π-allyl complexes. In $C_5H_5Mo(CO)_2(\pi\text{-allyl})$ and $C_5H_5Fe(CO)(\pi\text{-allyl})$, the π-allyl moiety can adopt two different configurations with respect to the rest of the molecule[61,62]. Whereas the isomeric forms of $C_5H_5Mo(CO)_2(\pi\text{-allyl})$, equivalent to compounds *38* and *39*, are transformed into each other, the isomers of $C_5H_5Fe(CO)(\pi\text{-allyl})$, like compounds *40*, do not rearrange[61,62].

In order to study substituent effects on the epimerization rate of square pyramidal compounds of the type $C_5H_5M(CO)_2LL'$, we synthesized the derivatives *41–49* by using instead of pyridine carbaldehyde-(2) the corresponding aldehydes for the condensation reaction with S-(−)-α-phenyl ethyl amine. In all the cases *41–49* the resulting pair of diastereoisomers could be separated into the optically active components a and b[63–65]; the a series is depicted in the formulas of Scheme 21.

The optically active square pyramidal complexes *39* and *41–49a* and b epimerize in solution, as shown in Fig. 4 for *38a* and b, some of them at lower, and some at higher temperatures[56,63–65]. The first-order decline in optical rotation yields overall rate constants k, which are the sum of the individual rate constants k_1 and k_2 for the interconversion of both epimers into each other[56,63–65] [Eq. (19)]. These first-order rate constants k for the approach to equilibrium were used for the calculation of the half-lives given in Table 1. A change in the solvent from DMF to toluene has only a small effect on the half-lives of the configurational stability[63].

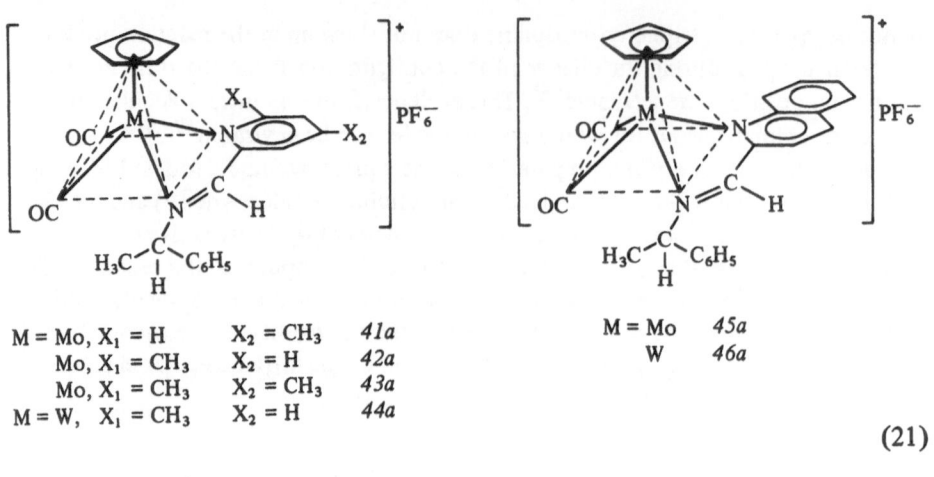

M = Mo, X₁ = H   X₂ = CH₃   *41a*
    Mo, X₁ = CH₃   X₂ = H   *42a*
    Mo, X₁ = CH₃   X₂ = CH₃   *43a*
M = W,   X₁ = CH₃   X₂ = H   *44a*

M = Mo   *45a*
W   *46a*

(21)

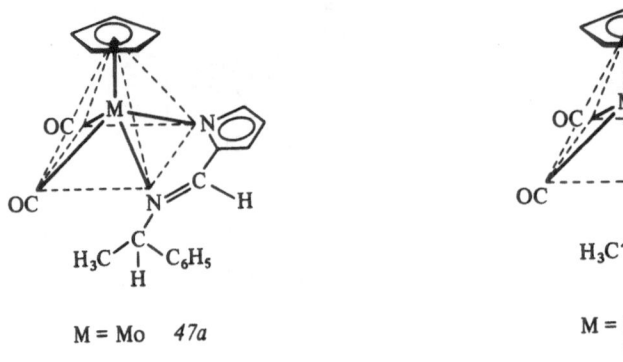

M = Mo   *47a*

M = Mo   *48a*
W   *49a*

Table 1. Half-lives [min] of the optical stability of the complexes *38, 39* and *41−49a* and *b* in DMF measured at 75 °C or extrapolated to 75 °C

Mo		W	
38a and *b*	26.3	*39a* and *b*	3.5
41a and *b*	35.1	*44a* and *b*	1110
42a and *b*	1670		
43a and *b*	2220		
38a and *b*	26.3	*39a* and *b*	3.5
45a and *b*	321	*46a* and *b*	178
47a and *b*	163		−
48a and *b*	0.07	*49a* and *b*	0.17

The upper half of Table 1 shows the half-lives of the unsubstituted complexes *38* and *39* and their derivatives with methyl groups in the o and/or p positions of the pyridine nucleus. It is obvious that a methyl group in the p position decreases the rate of epimerization a little whereas in the o position it decreases the rate of epimerization very much[64].

Let us take the relative rate constant

$$k_{rel} = k_X/k_H \cdot 100,$$ (22)

defined as the ratio of the rate constants for the substituted and the unsubstituted derivatives, and standardize it to 100. Then it can be seen that when 100 reference molecules of the unsubstituted Mo complex *38* epimerize only 75 molecules of its p-methyl derivative *41* and on average only 1.6 molecules of the o-methyl derivative *42* in the same time undergo the change in configuration. In the dimethyl derivative *43* the effects of the two methyl groups are additive[64] (Table 1).

The drastic decrease in the rate of epimerization of the o-substituted derivative *42* relative to the unsubstituted complex *38* and the p-substituted complex *41* is ascribed to the steric effect of the methyl group in the o position. In intramolecular isomerization, the steric hindrance should be greatly increased if a methyl group is introduced into the o position of the pyridine ring instead of an H atom[64]. Conversely, if the mechanism were dissociative, steric acceleration of the isomerization ought to be observed. These substituent effects therefore provide strong arguments in favor of an intramolecular epimerization reaction.

In addition to the methyl derivatives *41–44*, the epimerization of a number of other optically active Mo and W complexes was investigated polarimetrically. The lower half of Table 1 shows that in the quinoline derivatives *45* and *46* the rate of intramolecular epimerization is slowed down relative to the pyridine compounds *38* and *39*[64]. Obviously, an adjacent benzene ring increases the activation barrier for the pseudorotation in much the same way as an o-methyl group. On the other hand, the charge does not seem to have much influence on the epimerization rate, for the half-lives of the cationic pyridine derivative *38* and the neutral pyrrole derivatives *47* do not differ very much[56, 63]. If, however, the pyridine complex *38* and the corresponding phenyl derivative *48* are compared, a strong acceleration of the isomerization is observed in the case of phenyl. A correlation of corresponding Mo and W derivatives (Table 1) reveals that the W complexes in most cases epimerize a little faster than the Mo derivatives[56, 64]. Although these substituent effects support the formulation of the epimerization as an intramolecular reaction, further work is needed to clarify the critical dependence of the rate of isomerization on the nature of the chelate ring.

4.3. Other Racemization and Epimerization Reactions

Racemization reactions due to successive exchange of ligand roles and successive Walden inversions were discussed in Sections 3.1 and 3.2. The decarbonylation reaction seems to be stereospecific, yet, as mentioned in Section 3.3, subsequent to CO elimination there is a racemization of the decarbonylation products.

Attempts to cleave the metal-acyl and metal-ester bonds in $(-)$-$C_5H_5Fe(CO)$–$[P(C_6H_5)_3]COCH_3$ and $(-)$-$C_5H_5Fe(CO)[P(C_6H_5)_3]COOC_{10}H_{19}$ with I_2 led to optically inactive $C_5H_5Fe(CO)[P(C_6H_5)_3]I$[38]. It could, however, be shown that the iron-alkyl bond in 1.3-$(CH_3)(C_6H_5)C_5H_3Fe(CO)[P(C_6H_5)_3]CH_3$ can be cleaved under mild conditions by I_2, HI, and HgI_2 with some degree of stereospecificity[20, 21]. A suggested cause for the low stereospecificity of these reactions is the formation of a square pyramidal intermediate $[1.3$-$(CH_3)(C_6H_5)C_5H_3Fe(CO)[P(C_6H_5)_3](CH_3)I]^+I^-$, which could undergo rapid pseudorotation[21]. As regards the stereochemistry of the α-carbon atom, both retention and inversion have been evoked to account for the cleavage of metal-carbon bonds by halogens[20, 35, 70, 74].

Cobalt complexes of the type $C_5H_5Co(C_3F_7)(CNR)NCS$ were prepared in optically active form[67]. These complexes undergo a structural isomerization as far as the bonding of the thiocyanate ligand is concerned and an epimerization at the cobalt atom[67]. A photoracemization has been observed for (+)- and (−)-$C_5H_5Fe(CO)$-$[CNCH(CH_3)(C_6H_5)]I$[30]. In neither case are the mechanisms of the changes in configuration known.

5. Asymmetric Syntheses

(+)- and (−)-$C_5H_5Fe(CO)[P(C_6H_5)_3]CH_2OC_{10}H_{19}$ were used as the reagents for carbene transfer to the olefin *trans*-1-phenylpropene, with formation of the corresponding optically active cyclopropanes[13]. Similar results were obtained with an optically active chromium carbene complex containing an optically active phosphine[68].

The optically active molybdenum compounds *38* were used to determine the asymmetric induction in the ligands on reaction with organolithium compounds[69].

6. Conclusion

In this article we have described the different types of reactions that have been investigated by means of optically active compounds. It remains to determine the scope of the different mechanisms suggested, and to formulate stereochemical rules for use in further synthetic work.

7. References

[1] Brunner, H.: Angew. Chem. *81*, 395 (1969); Angew. Chem. int. Ed. Engl. *8*, 382 (1969).
[2] Brunner, H.: Ann. N. Y. Acad. Sci., *239*, 213 (1974).
[3] Brunner, H.: Angew. Chem. *83*, 274 (1971); Angew. Chem. int. Ed. Engl. *10*, 249 (1971).
[4] Brunner, H.: Chimia *25*, 284 (1971).
[5] Brunner, H.: Abstracts of Papers, Sixth International Conference on Organometallic Chemistry, Univ. of Mass., Amherst, Aug. 1973, P7.
[6] LaPlaca, S. J., Bernal, I., Brunner, H., Herrmann, W. A.: Angew. Chem., in press.
[7] Brunner, H., Schindler, H.-D.: Chem. Ber. *104*, 2467 (1971).
[8] Brunner, H., Schindler, H.-D.: Z. Naturforsch. *26b*, 1220 (1971).
[9] Kruck, T., Noack, M.: Chem. Ber. *97*, 1693 (1964).
[10] King, R. B., Bisnette, M., Fronzaglia, A.: J. Organometal. Chem. *5*, 341 (1966).
[11] Brunner, H., Schindler, H.-D.: J. Organometal. Chem. *24*, C7 (1970).
[12] Langer, M.: Diplomarbeit, TU München, 1971.

[13] Davison, A., Krusell, W. C., Michaelson, R. C.: J. Organometal. Chem. *72*, C7 (1974).
[14] Brunner, H., Langer, M.: J. Organometal. Chem. *54*, 221 (1973).
[15] Treichel, P. M., Shubkin, R. L.: Inorg. Chem. *6*, 1328 (1967).
[16] Darensbourg, M. Y.: J. Organometal. Chem. *38*, 133 (1972).
[17] Brunner, H., Langer, M.: J. Organometal. Chem. *87*, 223 (1975).
[18] Wojcicki, A.: Adv. Organometal. Chem. *12*, 31 (1974).
[19] Alexander, J. J., Wojcicki, A.: Inorg. Chim. Acta *5*, 655 (1971).
[20] Whitesides, G. M., Boschetto, D. J.: J. Amer. Chem. Soc. *93*, 1529 (1971).
[21] Attig, T. G., Wojcicki, A.: J. Amer. Chem. Soc. *96*, 262 (1974).
[22] Reich-Rohrwig, P., Wojcicki, A.: Inorg. Chem. *13*, 2457 (1974).
[23] Flood, T. C., Miles, D. L.: J. Amer. Chem. Soc. *95*, 6460 (1973).
[24] Brunner, H., Schmidt, E.: J. Organometal. Chem. *21*, P53 (1970).
[25] Brunner, H., Schmidt. E.: J. Organometal. Chem. *50*, 219 (1973).
[26] Seel, F.: Z. Anorg. Allg. Chem. *249*, 308 (1942).
[27] Brunner, H., Schmidt, E.: J. Organometal. Chem. *36*, C18 (1972).
[28] Brunner, H., Lappus, M.: Angew. Chem. *84*, 955 (1972); Angew. Chem. int. Ed. Engl. *11*, 923 (1972).
[29] Lappus, M.: Doktorarbeit. TU München, 1972.
[30] Brunner, H., Vogel, M.: J. Organometal. Chem. *35*, 169 (1972).
[31] Ugi, I., Fetzer, U., Eholzer, U., Knupfer, H., Offermann, K.: Neuere Methoden der präparativen organischen Chemie, Volume IV, p. 37. Weinheim: Verlag Chemie 1966.
[32] Wojcicki, A.: Adv. Organometal. Chem. *11*, 87 (1973).
[33] Calderazzo, F., Noack, K.: Coordination Chem. Rev. *1*, 118 (1966).
[34] Whitesides, G. M., Boschetto, D. J.: J. Amer. Chem. Soc. *91*, 4313 (1969).
[35] Johnson, R. W., Pearson, R. G.: Chem. Commun. *1970*, 986.
[36] Attig, T. G., Reich-Rohrwig, P., Wojcicki, A.: J. Organometal. Chem. *51*, C21 (1973).
[37] Su, S. R., Wojcicki, A.: J. Organometal. Chem. *27*, 231 (1971).
[38] Brunner, H., Strutz, J.: Z. Naturforsch. *29b*, 446 (1974).
[39] Alexander, J. J., Wojcicki, A.: Inorg. Chem. *12*, 74 (1973).
[40] Brunner, H., Aclasis, J., Langer, M., Steger, W.: Angew. Chem., *86*, 864 (1974); Angew. Chem. int. Ed. Engl. *13*, 810 (1974).
[41] Brunner, H., Aclasis, J.: J. Organometal. Chem., in press.
[42] Graham, M. A., Perutz, R. N., Poliakoff, M., Turner, J. J.: J. Organometal. Chem. *34*, C34 (1972).
[43] Wrighton, M.: Chem. Rev. *74*, 401 (1974).
[44] Bennett, M. J., Churchill, M. R., Gerloch, M., Mason, R.: Nature *201*, 1318 (1964).
[45] Chaiwasie, S., Fenn, R. H.: Acta Crystallogr. *24b*, 525 (1968).
[46] Barnett, K. W., Slocum, D. W.: J. Organometal. Chem. *44*, 1 (1972).
[47] Faller, J. W., Anderson, A. S.: J. Amer. Chem. Soc. *92*, 5852 (1970).
[48] Mawby, R. J., Wright, G.: J. Organometal. Chem. *21*, 169 (1970).
[49] Faller, J. W., Anderson, A. S., Jakubowski, A.: J. Organometal. Chem. *27*, C47 (1971).
[50] Kalck, P., Pince, R., Poilblanc, R., Russel, J.: J. Organometal. Chem. *24*, 445 (1970).
[51] Faller, J. W., Anderson, A. S., Chen, C.: J. Organometal. Chem. *17*, P7 (1969).
[52] Faller, J. W., Anderson, A. S., Chen, C.: Chem. Commun. *1969*, 719.
[53] King, R. B.: Inorg. Chem. *2*, 936 (1963).
[54] Brunner, H., Herrmann, W. A.: Chem. Ber. *105*, 3600 (1972).
[55] Brunner, H., Hermann, W. A.: Angew. Chem. *84*, 442 (1972); Angew. Chem. int. Ed. Engl. *11*, 418 (1972).
[56] Brunner, H., Herrmann, W. A.: Chem. Ber. *106*, 632 (1973).
[57] Gielen, M., Vanlauten, N.: Bull. Soc. Chim. Belges *79*, 679 (1970).
[58] Hässelbarth, W., Ruch, E.: Theoret. Chim. Acta *29*, 259 (1973).
[59] Musher, J. I., Agosta, W. C.: J. Amer. Chem. Soc. *96*, 1320 (1974).
[60] Brunner, H., Rambold, W.: J. Organometal. Chem. *64*, 373 (1974).
[61] Faller, J. W., Jakubowski, A.: J. Organometal. Chem. *31*, C75 (1971).
[62] Faller, J. W., Johnson, B. V., Dryja, T. P.: J. Organometal. Chem. *65*, 395 (1974).
[63] Brunner, H., Herrmann, W. A.: J. Organometal. Chem. *63*, 339 (1973).

64) Brunner, H., Herrmann, W. A.: J. Organometal. Chem. *74*, 423 (1974).

65) Brunner, H., Wachter, J.: J. Organometal. Chem. in press.

66) Frost, A. A., Pearson, R. G.: Kinetik und Mechanismen homogener chemischer Reaktionen p. 173. Weinheim: Verlag Chemie 1964.

67) Brunner, H., Rambold, W.: Z. Naturforsch. *29b*, 367 (1974).

68) Cooke, M. D., Fischer, E. O.: J. Organometal. Chem. *56*, 279 (1973).

69) Brunner, H., Wachter, J.: unpublished results.

70) Bock, P. L., Boschetto, D. J., Rasmussen, J. R., Demers, J. P. Whitesides, G. M.: J. Amer. Chem. Soc. *96*, 2814 (1974).

71) Attig, T. G., Wojcicki, A.: J. Organometal. Chem. *82*, 397 (1974).

72) Davison, A., Martinez, N.: J. Organometal. Chem. *74*, C 17 (1974).

73) Nicholas, K., Raghu, S., Rosenblum, M.: J. Organometal. Chem. *78*, 133 (1974).

74) Bock, P. L., Whitesides, G. M.: J. Amer. Chem. Soc. *96*, 2826 (1974).

Received August 5, 1975

Dynamics of Intramolecular Metal-Centered Rearrangement Reactions of Tris-Chelate Complexes

Louis H. Pignolet, Ph. D.

Department of Chemistry, University of Minnesota, Minneapolis, Minnesota 55455, U.S.A.

Contents

Abbreviations for ligand substituents, R:

benzyl	= Bz,
phenyl	= Ph,
ethyl	= Et,
methyl	= Me,
pyrrolidyl	= Pyr,
isopropyl	= iPr;

for complexes:

tris(N,N-disubstituted-dithiocarbamato)metal	= $M(R_1,R_2dtc)_3$,
tris(α-R-tropolonato)metal	= $M(RT)_3$,
tris(disubstituted-β-diketonato)-metal	= $M(R_1,R_2\text{-}\beta\text{-dik})_3$,
tris(oxalato)metal	= $M(OX)_3^{-3}$,
tris((+)-3-acetyl-camphorato)metal	= $M(atc)_3$;

for ligands:

1,10-phenanthroline	= phen.

I. Introduction

Since the time of Werner inorganic chemists have been intrigued by the observation of widely varying rates for isomerization reactions of metal complexes. Many studies have been carried out on reactions such as geometric isomerization and racemization especially with complexes which isomerize slowly, *i.e.*, with half-lives of several hours or longer. In this category are complexes of the traditionally non-labile metal ions such as Co(III), Cr(III), Rh(III), Ru(III), and Pt(II,IV). Much of this early work has been excellently reviewed by Basolo and Pearson.[1] The extent of these kinetically "slow" reactions was usually monitored by spectrophotometric or polarimetric means, although chromatographic techniques have been employed.[2] Since these techniques require isomer separation and/or resolution it was very tedious to carry out experiments. Recently nuclear magnetic resonance (NMR) spectroscopy has been used to measure rates of isomerization reactions. This technique greatly simplifies isomer detection[3] and with faster reactions often permits kinetic measurements without isomer separation. NMR line broadening techniques, (DNMR),[a] have now been extensively used to measure rates of reactions in which magnetically nonequivalent environments are averaged or interchanged on the NMR timescale.[4] Reactions with rates between ca. 10^{-2} to 10^6 sec^{-1} are best suited for DNMR studies and usually involve the kinetically "fast" metal ions such as Fe(III), Mn(III), V(III), Co(II), Ti(IV), Sc(III), Al(III) and Ga(III).

Several reviews have recently appeared which cover the topics of inter-[1, 6] and intramolecular[1, 5−11] rearrangement reactions of metal complexes. In order to minimize duplication this review will be limited to intramolecular metal-centered rearrangement reactions of six-coordinate tris-chelate complexes. Both kinetically "slow" (rates $\lesssim 10^{-2}$ sec^{-1}) and "fast" (rates $\gtrsim 10^{-2}$ sec^{-1})[b] complexes will be considered mainly from a mechanistic point of view. In addition various structural and electronic parameters of tris-chelate complexes will be scrutinized in order to assess their affect on the rate and mechanism of rearrangement.

Detailed mechanisms of intramolecular rearrangement reactions have been difficult to determine. Classical rate measurements seldom lead to unambiguous mechanistic predictions. Generally only after extensive examination of concentration, solvent, and substituent effects on the reaction rate can a general mechanistic class be proposed; for example, intra vs intermolecular or bond rupture of a bidentate chelate vs non-bond rupture twist pathways. Indeed, only two examples of "slow" complexes are known where detailed rate comparisons for geometrical and optical isomerizations were made and used to eliminate several mechanisms; however, a single most probable pathway was not demonstrated in either case.[12, 13] Only with DNMR can detailed environmental site interchanges be directly observed and with this in-

---

a) The term DNMR, Dynamic Nuclear Magnetic Resonance, refers to the process of recording and usually computer simulating exchange broadened NMR spectra at a number of temperatures in order to determine mechanistic and/or kinetic information.

b) The terms "fast" and "slow" are used rather than the more familiar labile and nonlabile because the latter imply an intermolecular reaction.

formation the most probable mechanism can sometimes be deduced. Even with this information one can only determine the most reasonable mechanism which gives the observed site interchanges. The DNMR experiment unambiguously gives the site interchange pattern (resolution permitting) and the scientist must then hypothesize a consistent mechanism(s) based on chemical intution. For example, a proposed transition state should be a reasonable geometry for the rearranging molecule. Often the rule of "least motion" is invoked. It is well to keep in mind that rearrangement mechanisms are necessarily hypothetical and that only a particular site interchange or ligand permutational reaction[c] can be determined by DNMR. In this review the various physically reasonable mechanisms will be described first and then related to their respective ligand permutational reactions or rearrangement modes.[14], [d]

Of all the six-coordinate chelate complexes studied by DNMR to date unique rearrangement modes have been determined for only two types of tris(bidentate) chelates, tris-(N,N-disubstituted dithiocarbamates), *1*, and tris-(α-substituted tropolonates), *2*.[e] These complexes in addition to many others where unique modes have not been established will be discussed in part IV.

1 M(R$_1$, R$_2$dtc)$_3$ *2* M(RT)$_3$

II. Nuclear Magnetic Resonance Determination of Rates and Mechanisms (DNMR)

For rearrangement reactions of the kinetically "fast" variety, *i.e.*, for complexes which are stereochemically nonrigid, DNMR techniques have in several cases permitted detailed mechanistic conclusions via the observation of site interchanges.[5−8] The first experiments were carried out by Fay and Piper[15] on tris(trifluoroacetylacetonato) complexes of aluminum(III), gallium(III) and indium(III), *3*, using ^{19}F DNMR. These complexes which are members of the class of octahedral tris chelates M(A−B)$_3$

M(CF$_3$, CH$_3$-β-dik)$_3$

3

---

[c] A ligand permutational reaction refers to a specific permutation of indexed ligand positions. For a six coordinate complex there are 6! or 720 such permutations including the identity, however, fortunately all do not yield distinguishable isomers.

[d] Musher has adopted the phrase rearrangement mode which refers to a set of physically equivalent or indistinguishable ligand permutational reactions (see Section III.B).

[e] Ligand abbreviations are given on p. 92.

where A—B represents an unsymmetrical bidentate ligand can exist as cis and trans geometric isomers each of which is enantiomeric, *4*. The notation used in *4* and

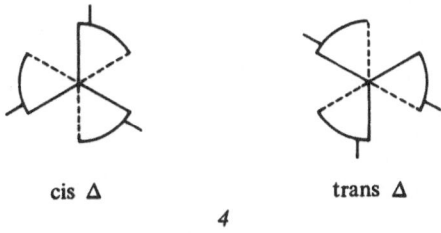

cis Δ   trans Δ

4

throughout this review depicts an unsymmetrical chelate ligand as a curved line with a tag at one end. In the case of the Ga complex four ^{19}F resonances are observed below ca. 49 °C which result from the cis (one resonance, C_3 symmetry) and trans (three resonances, C_1 symmetry) isomers. As the temperature is increased the four lines simultaneously collapse into one sharp resonance at 96 °C as shown in Fig. 1a.[15] Kinetic parameters for cis-trans isomerization can be determined by computer simulation of the exchange broadened resonances using one of several

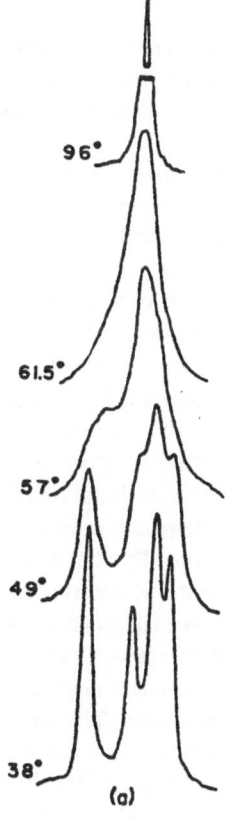

96°

61.5°

57°

49°

38°

(a)

Fig. 1. a) Fluorine DNMR spectrum of Ga(CF$_3$, CH$_3$-β-dik)$_3$. From Ref.[15]

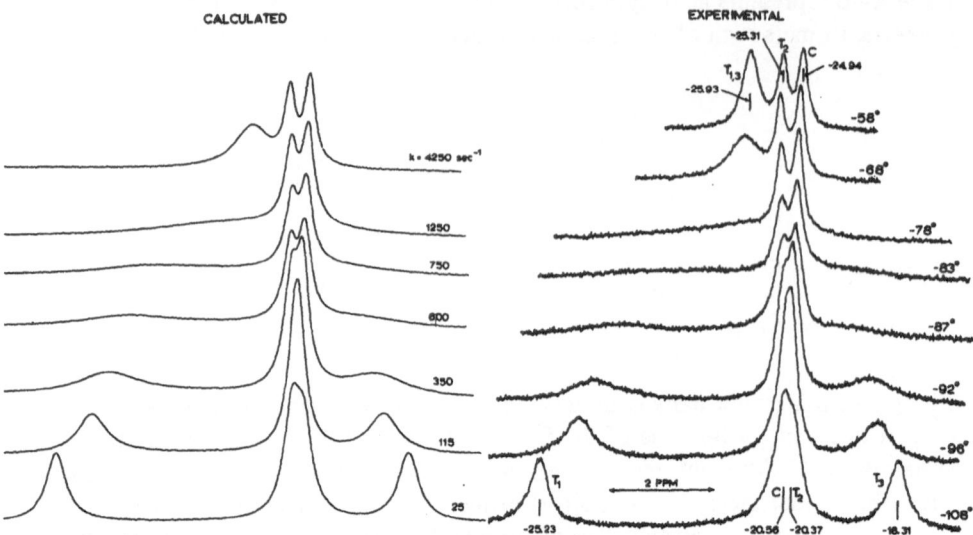

Fig. 1. b) Proton DNMR of N—CH$_3$ region of Fe(CH$_3$, Phdtc)$_3$ at 100 MHz in CD$_2$Cl$_2$ solution. observed and calculated line shapes are shown. From Ref.[19]

computer programs.[16] DNMR kinetic techniques have been described in detail elsewhere and recently several very complex studies have been carried out.[17, 18]

It is also possible to ascertain the mode of a rearrangement reaction by careful examination of the coalescence pattern. The various rearrangement modes (see Section III.B) usually lead to different site interchange patterns which depend on the ligand substituents. In the above case a computer simulation was not performed, however, if the four sites were found to coalesce in a specific non-random way some mechanistic conclusions possibly could be made. Even without careful analysis, however, several rearrangement mechanisms were eliminated as sole pathways.[15] These results should be contrasted with those of Fe(CH$_3$, Phdtc)$_3$, another M(A—B)$_3$ type complex whose N—CH$_3$ DNMR spectra are shown in Fig. 1b.[19] In this case only two of the four sites are interchanged in the temperature interval shown and the calculated spectra illustrate the result of a kinetic line shape analysis. The unequal cis and trans populations permitted assignment of the cis methyl resonance, C. Only two rearrangement modes or permutational sets (see Section III.B) are consistent with the T$_1$ ⇄ T$_3$ site interchange and only one of these predicts enantiomerization (Λ ⇄ Δ metal centered inversion). Further experiments[19] with unsymmetrical dtc ligands which contain diastereotopic protons (*vide infra*) were able to show that enantiomerization occurs and have lead to an unambiguous assignment of the sole rearrangement mode which is illustrated in 5 for the trans Δ isomer. Note that cis ⇄ trans isomerization does not accompany enantiomerization while two of the trans methyl sites, X and Z, are interchanged. The letters refer to magnetic environments or sites and the numbers label the methyl groups. The most reasonable mechanism which gives this mode is the trigonal or Bailar[20] twist (see Section III.A).

The above examples illustrate how kinetic and mechanistic information can be extracted from DNMR experiments. Specific site interchange patterns are most use-

trans Δ   trans Λ

5

ful for the latter, however, it is also desirable to measure rates of overall reactions such as geometric isomerization and enantiomerization. Significant mechanistic information can be derived from ratios of the various microscopic rate constants for such reactions. Indeed two independent studies[12, 13] on kinetically "slow" tris(un-symmetrical-β-diketonate) complexes of Co(III) in which the various geometric and optical isomers were completely or partially separated and subjected to kinetic analysis have yielded similar mechanistic conclusions (see Section IV.C). These experiments were fruitful because the various rearrangement modes (Section III) lead to different ratios of microscopic rate constants.

An important DNMR technique for measuring rates of enantiomerization reactions of "fast" complexes is the incorporation of diastereotopic groups[21] into the ligand framework. Diastereotopic groups as used here are substituents which have nuclei which are nonequivalent solely due to the dissymmetry about the metal center and their occurrence is best illustrated by an example. Figure 2[22] shows the methyl DNMR spectrum of Al(iPr, iPr-β-dik)$_3$, a M(A*—A*)$_3$ type complex where the asterisks indicate diastereotopic lables. The isopropyl methyl groups λ and δ are diastereotopic and hence two spin-spin doublets due to coupling with the methine proton are observed at 37 °C. As the temperature is increased the two doublets coalesce into one due to rapid enantiomerization of the complex. A total line shape analysis has been carried out.[23] The use of diastereotopic groups is an extremely convenient method for measuring enantiomerization rates of kinetically "fast" complexes.[5—8]

Line shape analysis of coalescing NMR peaks gives a pre-exchange lifetime, τ, at each temperature where significant exchange broadening occurs. For intramolecular processes the reciprocal of the lifetime equals the first order rate constant k. The usual work-up of the data consists of a standard Erying or Arrhenius plot to give the enthalpy and entropy of activation, $\Delta H^{\ddagger}$ and $\Delta S^{\ddagger}$, or the activation energy and frequency factor, E_a and $\log A$, respectively. A general observation which results from the many DNMR kinetic studies reported is that for intramolecular processes $\Delta S^{\ddagger}$ is usually near zero ($\log A = 13.2$) and almost always within 15. e.u. of zero. Values of $\Delta S^{\ddagger}$ outside of this range should be viewed with caution if other factors indicate intramolecularity. Another rule of thumb which is often extremely useful is the realization that for a simple two site exchange the rate constant at the temperature of maximum line broadening or coalescence is approximately equal to the peak separation in Hz in the absence of exchange. Hence a very large chemical shift difference between two exchanging environments permits the measurement of rather

97

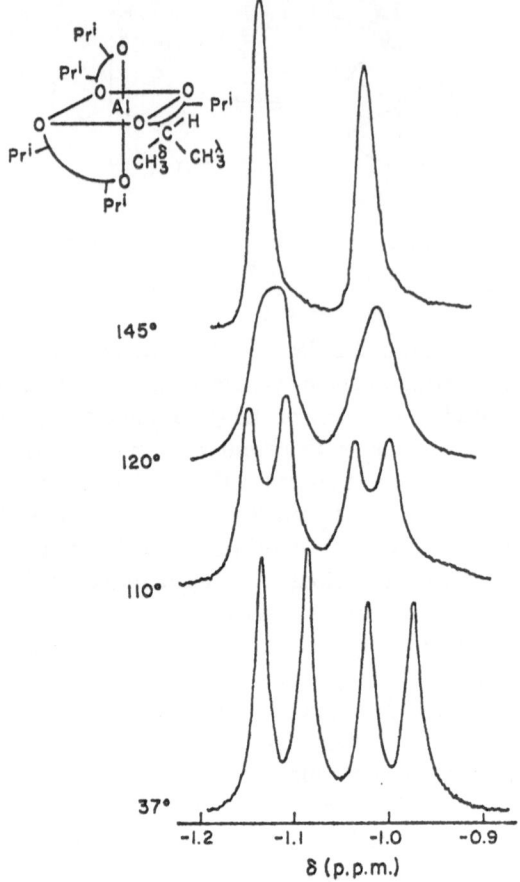

Fig. 2. DNMR spectrum of the isopropyl methyl groups of Al(iPr, iPr-β-dik)$_3$ in chlorobenzene solution. From Ref.[22]

fast rates. Rate constants as large as 10^6 sec^{-1} have been measured[24] by DNMR where the exchanging environments were near paramagnetic metal centers and thus experienced large isotropic shifts.[25]

III. Rearrangement Pathways

The previous section outlined how rates and modes of rearrangement reactions can be determined by DNMR techniques. This section will describe in some detail the various intramolecular rearrangement mechanisms and modes which tris chelate complexes can undergo. A discussion of mechanisms must, of course, include a consideration of the experimental techniques used. Classically for kinetically "slow" complexes the various geometric and optical isomers are separated and their kinetics measured by spectrophotometric or chromatographic means. Hence for each mechanism we must derive the isomerization rate relations between the various isomers. This has been done in detail for the $M(A-B)_3$ case.[12, 13]

DNMR gives a site interchange pattern and the rate relations for the various interchanges. The types of complexes which have been studied are

$$M(A^*-A^*)_3, \quad M(A-B)_3, \quad M(A-B^*)_3, \quad M(A-A)_2(B-B), \quad M(A^*-A^*)_2(B-B),$$
$$M(A-B)_2(C-C),$$

and infrequently $M(A-B^*)_2(C-C)$, and $M(A-A)_2(C^*-C^*)$. It would be a formidable task to enumerate all of the site interchanges and relative rates in each of these cases for each rearrangement mode, therefore, the details will be shown for the $M(A-B^*)_3$ and $M(A-B)_2(C-C)$ cases only. It is hoped that these examples will enable the reader to carry out the analysis for any of the other cases.

Metal centered rearrangement reactions will be reviewed by two general approaches. First (part A) we will consider only the rearrangements which are physically reasonable and therefore a description of the most likely geometry of the transition state will be included. These rearrangements will be referred to as mechanisms, for example, twists via trigonal prismatic or bond ruptures via trigonal bipyramidal transition states. Secondly (part B) we will consider the analyses by permutational reactions as recently carried out by Musher[14] and Eaton et al.[26, 27] This method has the advantage of assuring that all rearrangements have been considered, however, it only considers permutations of sites and not motions of ligands. It is important to realize that DNMR can only give information about site interchanges and hence only a class of permutational reaction or rearrangement mode can be proven. The assignment of a mechanism will always be hypothetical and realistically requires added chemical knowledge.

A. Mechanisms *via* Idealized Transition States

Intramolecular metal centered rearrangement reactions for tris chelate complexes result in two types of stereochemical change:

enantiomerization ($\Lambda \rightleftarrows \Delta$) and

isomerization (cis $\rightleftarrows$ trans).

Mechanisms for such reactions have been viewed by the effective coordination number of the transition state. Bond rupture processes proceed via five-coordinate transition states for which idealized square-pyramidal (SP) and trigonal-bipyramidal (TBP) geometries with the dangling ligand axial or equatorial have been considered. Twist mechanisms are considered to proceed via six-coordinate transition states with idealized trigonal prismatic (TP) geometry. Twists of the chelate rings about the real or pseudo C_3 axis in the case of the cis or trans isomer, respectively, and about the imaginary C_3 axes have been considered.[28] These non-bond-breaking twist mechanisms have been referred to as trigonal or Bailar[20] twists (r-C_3 or p-C_3) and rhombic or Rây-Dutt[29] twists (i-C_3). Examples are discussed below for the $M(A-B)_2C-C$ case. The various mechanisms often lead to different stereochemical results, for example in the $M(A-B)_2(C-C)$ case, bond rupture through a TBP-equatorial transition state leads to isomerization but not enantiomerization whereas a TBP-axial transition state leads to isomerization with enantiomerization. All twist processes result in enantiomerization. Several bond rupture mechanisms are illustrated in Fig. 3.

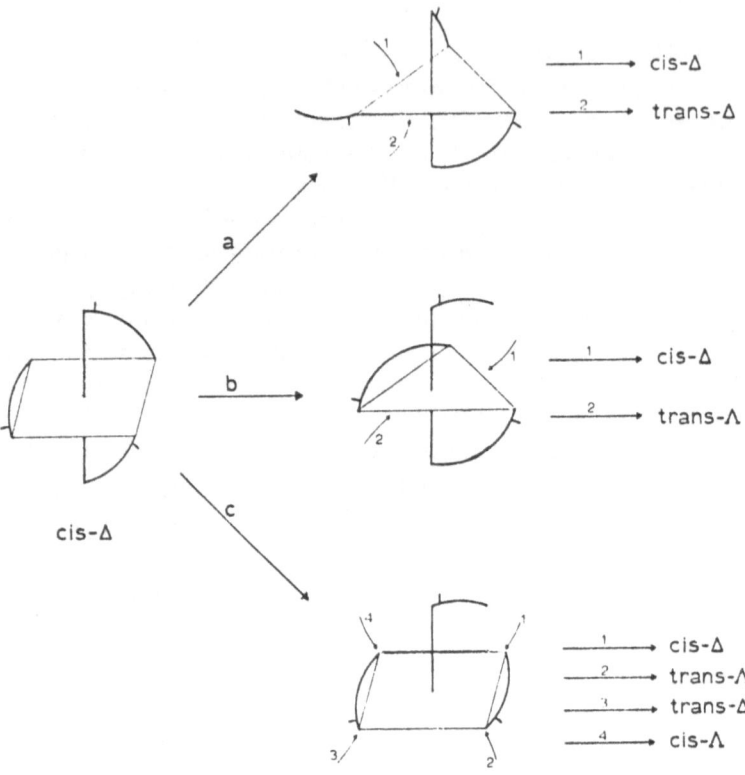

Fig. 3. Stereochemical rearrangement of the cis-Δ isomer of a $M(A-B)_3$ complex via a bond rupture mechanism proceeding through (a) a trigonal bipyramidal intermediate with a dangling equatorial ligand, (b) a trigonal bipyramidal intermediate with a dangling axial ligand, and (c) a square pyramidal intermediate with a dangling axial ligand. The numbers label possible sites for reattachment of the end of the ligand. From Ref.[6]

The stereochemical consequences of these rearrangement mechanisms have been worked out in detail for the $M(A-B)_3$[12] and $M(A-B)_2(C-C)$[30] cases and are summarized in topological correlation diagrams.[31] For the latter case three isomers are possible each of which is enantiomeric, 6. Several of the possible twist mechanisms

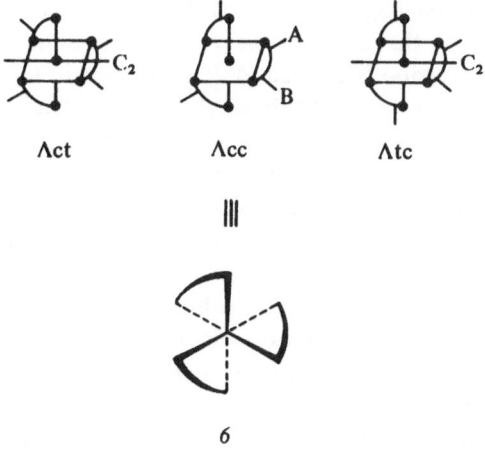

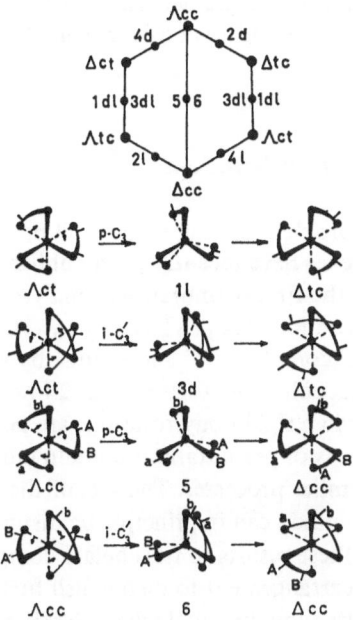

Fig. 4. Topological correlation diagram for the interconversion of $M(A-B)_2(C-C)$ isomers by a twist mechanism. Twist motions about the $p-C_3$ and one of the $i-C_3$ axes of Λ_{ct} and Λ_{cc} are illustrated. From Ref.[30]

and the topological correlation diagram which results from all possible twists are illustrated in Fig. 4.[30] In the correlation diagram all possible isomer interconversions are summarized. Isomers are located at the apices and transition states at the midpoint of edges and the diagonal.[f] Several stereochemical results which are apparent in the diagram are that all reactions result in enantiomerization and that direct ct $\rightleftarrows$ tc interconversion is allowed. A consideration of correlation diagrams for the bond rupture mechanisms as well shows all of the constraints which are placed on the one-step interconversions. This kind of information has permitted mechanistic conclusions on $M(A-B)_3$ systems of "slow" Co(III) β-diketonate complexes by kinetic analysis[12, 13] and on "fast" bid(dithiocarbamato)dithiolene iron complexes by DNMR.[30, 32] With the latter complexes site interchanges were also used to elucidate the mechanism. For example, the $\Lambda_{cc} \rightarrow \Delta_{cc}$ interconversion shown in Fig. 4 has a different site interchange pattern for substituents labeled A, B, a, and b depending on whether the interconversion proceeds via transition state 5 or 6. The former (p-C_3 twist) averages nuclei a and b but not A and B whereas in the latter (i-C_3') both pairs of nuclei a↔b and A↔B, are averaged.

In the above illustrations the mechanisms discussed are considered to be physically reasonable and accordingly involve transition state geometries which are known for other metal complexes. This approach in no way guarantees that all rearrangements have been considered although in every case studied thus far by DNMR one or several of the above rearrangements do account for the observed site interchange patterns. However, it is most useful and necessary to mathematically enumerate all possible permutational reactions and to consider their stereochemical and site interchange consequences.

B. Permutational Rearrangement Reactions

Several approaches to the problem of enumerating all possible permutational reactions of tris chelate complexes have recently been published.[14, 26, 27]

Musher[14] carried out the first permutational analysis on the general class of six-coordinate octahedral complex $ML_1 L_2 L_3 L_4 L_5 L_6$. The analysis is similar to the one on five coordinate TBP complexes[33] and is done by writing out the 6! = 720 permutations of the indices; collecting together the 24 permutations which give indistinguishable isomers when rigid-body rotations are permitted; and observing which of the 30 resultant sets of permutations contain individual permutations corresponding to physically similar processes. The sets of these sets are called modes of rearrangement[14] and each mode can in principle be distinguished by DNMR experiments. Musher[14] applied these results to tris chelate complexes which necessitated the use of split modes of rearrangement to distinguish further the "types" of stereochemical change. The results show in detail what DNMR experiments are needed in order to establish a class or mode of rearrangement. Musher[14] also pointed out for the first time the ambiguities associated with several previous DNMR experiments

---

[f] The following transition states result from twists about the indicated axes of the Λ isomers: Λ_{ct}: p-C_3, 1l; i-C_3, 4l(2); i-C_3', 3d; Λ_{tc}: p-C_3, 1l; i-C_3, 2l(2); i-C_3', 3l; Λ_{cc}: p-C_3, 5; i-C_3, 2d; i-C_3', 6; i-C_3'', 4d. Axes are defined in an analogous manner for the three isomers. Note that two of the imaginary axes of ct and tc are equivalent. Transition states 1l, 3d, 5, and 6 are illustrated in Fig. 4 and the remainder may be deduced from the correlation diagram in this figure.

(see Section IV). An equivalent but mathematically more complex analysis was carried out by Eaton and Eaton.[26] In this review the results of Eaton and Eaton[26] will be used because the configurational changes and site interchanges are more clearly tabulated.

Eaton et al.[26, 27] applied the method of molecular symmetry groups for non-rigid molecules introduced by Longuet-Higgins[34] to complexes of the type $M(A-B)_3$ and $M(A-B)_2(C-C)$ and derived averaging sets of NMR indistinguishable permutations. The resulting site interchanges which result from each averaging set can then be derived along with any configurational change. For example, the analysis for the $M(A-B*)_3$ case is as follows. The notation used to indicate nonequivalent environments or sites is shown in 7 for a T isomer where the magnetic environments x, y and z are, in principle, distinguishable for any unsymmetrical ligand whereas the environments r and s apply only to diastereotopic substituents. Several permutations are illustrated in 8.[26] The cis isomer 8a has the notation [135−462][g] and permuta-

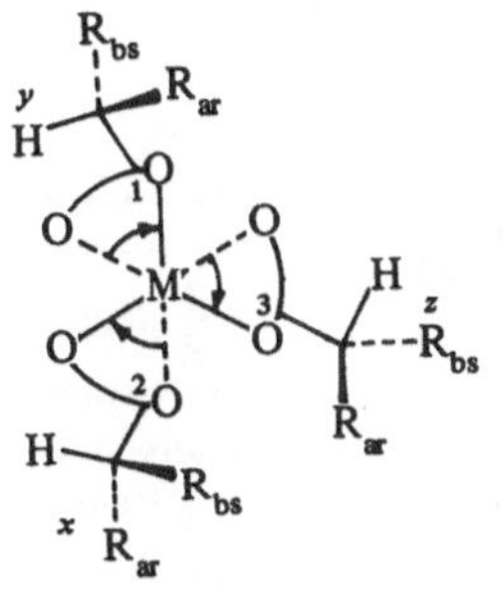

7

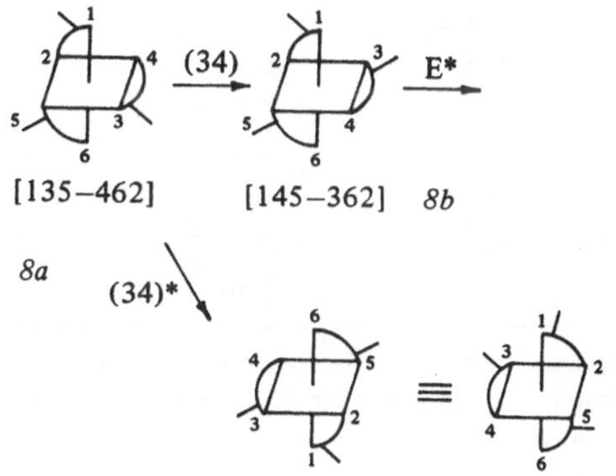

[135−462]   [145−362]   8b

8a

(34)*

8c   [154−263]

8

---

tion (34)[h] gives the trans isomer *8b*. The permutation E*, the inversion of all ligands through the center of mass, on *8b* gives *8c* which is obtained from *8a* by permutation-inversion (34)*. The set of all permutations and permutation-inversions is a group of order 384 which factors into a group of order 16 consisting of the rearrangements among isomers not related by rigid-body rotations and a group of order 24 consisting of the rigid-body rotations of the molecule. The 16 operations in the group of order 16 constitute the complete group of rearrangements of stereochemically nonrigid tris chelate complexes.[26] The effect of each of the 16 operations on the isomer [135–462] is given in Table 1.[26] Permutations which give the same net DNMR averaging

Table 1.[26] Permutational analysis of M(A–B)$_3$ NMR spectra

Operation	Resulting isomer[1])	Averaging set	Net configurational change	Net site interchanges (trans)
E	[135–462]	A$_1$	None	None
(12) (34) (56)	[153–642]	A$_2$	None	(yz)
(12)	[164–532]		Cis → trans	
(34)	[145–362]	A$_3$	Trans → $\frac{1}{3}$ cis	(xy), (xz)
(56)	[136–452]		+ $\frac{2}{3}$ trans	
(12) (34)	[163–542]		Cis → trans	
(34) (56)	[146–352]	A$_4$	Trans → $\frac{1}{3}$ cis	(xzy), (xyz)
(12) (56)	[154–632]		+ $\frac{2}{3}$ trans	
E*	[153–264]	A$_5$	$\Delta \rightleftarrows \Lambda$	(rs)
(12) (34) (56)*	[135–246]	A$_6$	$\Delta \rightleftarrows \Lambda$	(yz), (rs)
(12)*	[146–235]		$\Delta \rightleftarrows \Lambda$	
(34)*	[154–263]	A$_7$	Cis → trans	(xy), (xz), (rs)
(56)*	[163–254]		Trans → $\frac{1}{3}$ cis	
			+ $\frac{2}{3}$ trans	
(12) (34)*	[136–245]		$\Delta \rightleftarrows \Lambda$	
(34) (56)*	[164–253]	A$_8$	Cis → trans	(xyz), (xzy)
(12) (56)*	[145–236]		trans → $\frac{1}{3}$ cis	(rs)
			+ $\frac{2}{3}$ trans	

[1]) Isomers resulting from performing the indicated operations on [135–462] are given as examples. Columns 3, 4, and 5 summarize net effects of operations on all 16 isomers.

patterns for the nonequivalent sites of 7 are combined into averaging sets A$_1$ through A$_8$.

The DNMR experiment with a M(A–B*)$_3$ type complex can at best distinguish between the eight averaging sets. Indeed all averaging sets with the exception of the pairs (A$_3$, A$_4$), (A$_5$, A$_6$) and (A$_7$, A$_8$) are distinguishable from changes in signal

---

[h] Standard permutation notation and means 3 replaces 4 and 4 replaces 3.

multiplicity alone. In order to make distinctions within the pairs a careful lineshape fit is necessary. In the case of $M(A-B)_3$ where enantiomerization cannot be sensed the averaging sets are pairwise indistinguishable from changes in signal multiplicity alone: (A_1, A_5), (A_2, A_6), (A_3, A_7) and (A_4, A_8).[26] The latter two pairs may be distinguished via lineshape analysis. Both pairs effect cis-trans isomerization and thus average all four signals to one but with slightly different coalescence patterns.

The averaging sets can be correlated with rearrangement mechanisms (Section III.A) which give the same site interchange as follows: p-C_3 or r-C_3 twists, A_6; i-C_3 twists, A_8; TBP-axial, A_7; TBP-equatorial, A_3; SP-axial, $A_3 + A_6 + A_8$.

A similar analysis was also carried out on the $M(A-B)_2(C-C)$ type complex.

Table 2.[26] Permutational analysis of $M(A-B)_2(C-C)$ NMR spectra

Operation	Averaging set	Net configurational change	Net site interchanges (cc isomer)	$M(A-B)_3$ analog
E	A_1'	None	None	A_1
(12) (34) (56)	A_2'	ct $\rightleftarrows$ tc	(mn)	A_2
(56)	A_3'	None	(fg)	A_3
(12)	A_4'			A_3
(34)				A_3
(34) (56)	A_5'	tc $\rightleftarrows$ cc $\rightleftarrows$ ct		A_4
(12) (56)				A_4
	A_6'			
(12) (34)	A_7'	ct $\rightleftarrows$ tc	(fg), (mn)	A_4
E*		$\Delta \rightleftarrows \Lambda$	(rs)	A_5
	A_8'			
(12) (34) (56)*		ct $\rightleftarrows$ tc, $\Delta \rightleftarrows \Lambda$	(mn), (rs)	A_6
(56)*	A_9'	$\Delta \rightleftarrows \Lambda$	(fg), (rs)	A_7
(12)*				A_7
(34)*		tc $\rightleftarrows$ cc $\rightleftarrows$ ct		A_7
(34) (56)*		$\Delta \rightleftarrows \Lambda$	(rs)	A_8
(12) (56)*				A_8
(12) (34)*	A_{10}'	ct $\rightleftarrows$ tc, $\Delta \rightleftarrows \Lambda$	(fg), (mn), (rs)	A_8

Results are given in Table 2.[26] In this case there are three geometrical isomers 6 and a total of four magnetic environments shown in 9.[26] The ct and tc isomers have C_2 symmetry so each possesses a single environment for Substitutents A, B and C whereas the cc isomer has two C sites, f and g, and two B sites, m and n, where B is arbitrarily represented by the flag of chelate A–B. The notation r, s refers to diastereotopic environments on either the A–B or C–C chelate. Table 3 shows changes in signal multiplicity which result from the ten averaging sets. These results will be referred to in Section IV.D. The most information from DNMR experiments with regard to $M(A-B)_2(C-C)$ type complexes is obtained by comparing the NMR averaging pattern for a complex with nondiasterotopic substituents on both

A–B and C–C with a closely related complex containing diastereotopic groups on C–C.

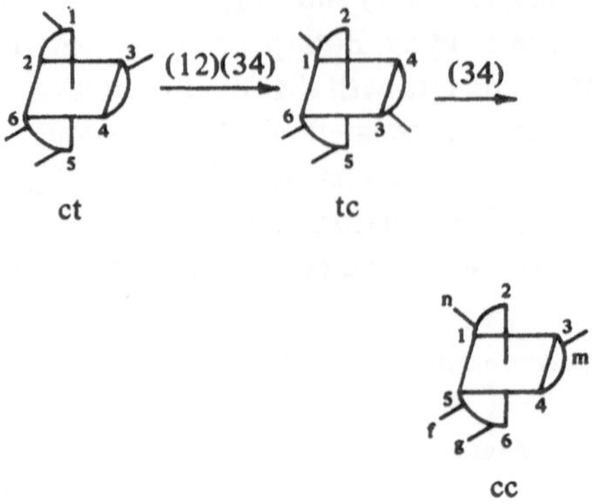

ct   tc

cc

9

Table 3.[26] Changes in signal multiplicity for diastereotopic and nondiastereotopic ligands

	A–B		C–C	
	Nondiastereotopic	Diastereotopic	Nondiastereotopic	Diastereotopic
A_1'	$4 \to 4$	$8 \to 8$	$4 \to 4$	$8 \to 8$
A_2'	$4 \to 2$	$8 \to 4$	$4 \to 3$	$8 \to 6$
A_3'	$4 \to 4$	$8 \to 8$	$4 \to 3$	$8 \to 6$
A_4'	$4 \to 1$	$8 \to 2$	$4 \to 1$	$8 \to 2$
A_5'	$4 \to 2$	$8 \to 4$	$4 \to 2$	$8 \to 4$
A_6'	$4 \to 4$	$8 \to 4$	$4 \to 4$	$8 \to 4$
A_7'	$4 \to 2$	$8 \to 4$	$4 \to 3$	$8 \to 4$
A_8'	$4 \to 4$	$8 \to 4$	$4 \to 3$	$8 \to 4$
A_9'	$4 \to 1$	$8 \to 2$	$4 \to 1$	$8 \to 2$
A_{10}'	$4 \to 2$	$8 \to 4$	$4 \to 2$	$8 \to 4$

IV. Experimental Studies

A Tris(dithiocarbamates)

This class of paramagnetic tris chelate complex, *1*, has been extensively studied by DNMR by Pignolet and co-workers[19, 35–39] and although these compounds were not the first examined by DNMR they do represent the largest collection for which a unique rearrangement mode or averaging set has been demonstrated. The first compound examined was $Fe(CH_3, Phdtc)_3$[19, 35] a compound whose NMR spectrum had previously been reported by others[40] but whose stereochemical nonrigidity was not recognized. The DNMR spectra of the $N-CH_3$ region is shown in Fig. 1b[19] along with the calculated line shapes. At $-108 °C$ the complex is rigid and the four resonances result from the C and T isomers (see Section II). The C resonance can be identified by signal integration via computer calculation and by its dynamic behavior. As the temperature is increased two of the trans environments, T_1 and T_3, are interchanged while the C and remaining T_2 environments are unaffected. This specific site interchange for the T isomer, and the absence of geometric isomerization ($C \rightleftarrows T$) are consistent only with averaging sets A_2 and A_6 (see Table 1). Of these only set A_6 requires that enantiomerization accompany the site interchange. NMR investigation of the similar compound $Fe(Bz, Bzdtc)_3$[19] where Bz = benzyl showed that averaging of the diastereotopic methylene environments occurred with similar activation parameters. This result implies but does not prove that enantiomerization and the (yz) site interchange occur simultaneously for $Fe(dtc)_3$ complexes which therefore must rearrange via averaging set A_6. Further experiments with $Fe(CH_3, -Bzdtc)_3$[19] showed conclusively that averaging set A_6 is the primary rearrangement mode for the T isomer and most probably for the C isomer as well, although site interchanges cannot be observed in the latter.

The $Fe(dtc)_3$ complexes also exhibit a high temperature process which averages the remaining three CH_3 resonances of the (CH_3, Ph) substituted complex. This process results in geometric CH_3 isomerization and has been assigned to S_2C-N bond rotation[41] although several metal centered rearrangements, for example, averaging sets A_3, A_4, A_7 and A_8, will also account for the results.

The low temperature process is obviously intramolecular because all intermolecular rearrangements would result in geometric isomerization. Ligand exchange has been observed in these complexes and occurs at a much slower rate than either the low or high temperature processes so the high temperature process is intramolecular as well.

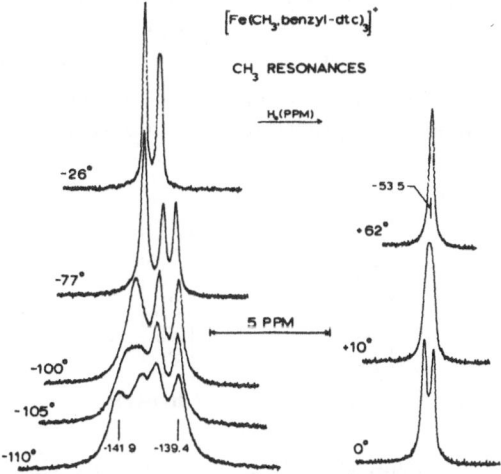

Fig. 5. DNMR spectrum of the $N-CH_3$ resonances of $[Fe(CH_3, Bzdtc)_3]BF_4$ in CD_2Cl_2 solution. From Ref.[36]

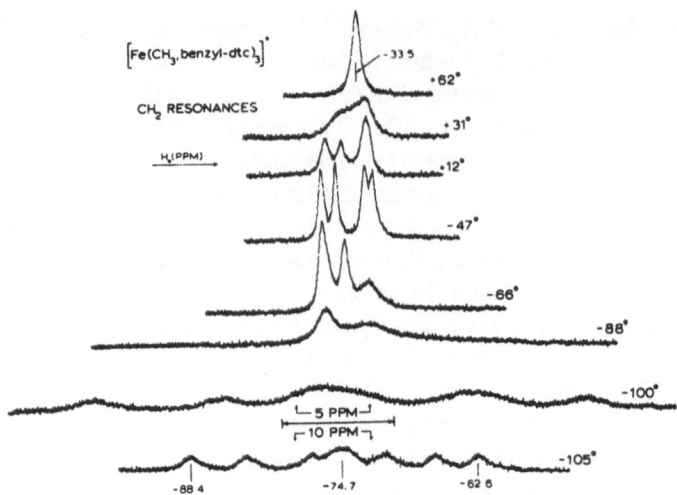

Fig. 6. DNMR spectrum of the N–CH₂ resonances of [Fe(CH₃, Bzdtc)₃]BF₄ in CD₂Cl₂ solution. From Ref.[36]

DNMR experiments have also been carried out with [Fe(R′dtc)₃]BF₄ complexes[19, 36] where the formal oxidation state of iron is +4. These complexes can be prepared by air oxidation of the corresponding iron(III) complexes in the presence of BF₃[42] or by electrochemical means.[43] The DNMR spectra of [Fe(CH₃,-Bzdtc)₃]BF₄ is shown in Figs. 5 and 6[36] for the N–CH₃ and N–CH₂ groups, respectively. The N–CH₃ coalescence pattern for the low temperature process (−77 to −110 °C) is the same as observed for Fe(CH₃, Phdtc)₃ and therefore is consistent only with averaging set A₂ or A₆ both of which give the (yz) site interchange for the T isomer. An analysis of the coalescence pattern of the N–CH₂ group reveals that only averaging set A₆ is operative. This averaging set which is that given by the trigonal twist mechanism predicts that eight lines all coalesce simultaneously via four pairwise interchanges into four resonances (Fig. 6, LTP). This pattern can be derived from the rearrangement reactions shown in Fig. 7[36] for the C and T isomers. The site interchange pattern which results using the notation in Fig. 7 is given by *10* for the T isomer where numbers and primed numbers represent the methylene nuclei

$$
\begin{bmatrix}
1\text{-}a \\
1'\text{-}b \\
2\text{-}c \\
2'\text{-}d \\
3\text{-}e \\
3'\text{-}f
\end{bmatrix}
\xrightleftharpoons{\text{pseudo } C_3}
\begin{bmatrix}
1\text{-}f \\
1'\text{-}e \\
2\text{-}d \\
2'\text{-}c \\
3\text{-}b \\
3'\text{-}a
\end{bmatrix}
\qquad\qquad 10
$$

$$\text{trans } \Lambda \qquad\qquad\qquad \text{trans } \Delta$$

and letters refer to the nonequivalent magnetic environments. The four site interchanges are: (af), (be), (cd) for the T isomer and (ab) for the C isomer. The DNMR spectra in Fig. 6 are in complete agreement with this pattern. The eventual coales-

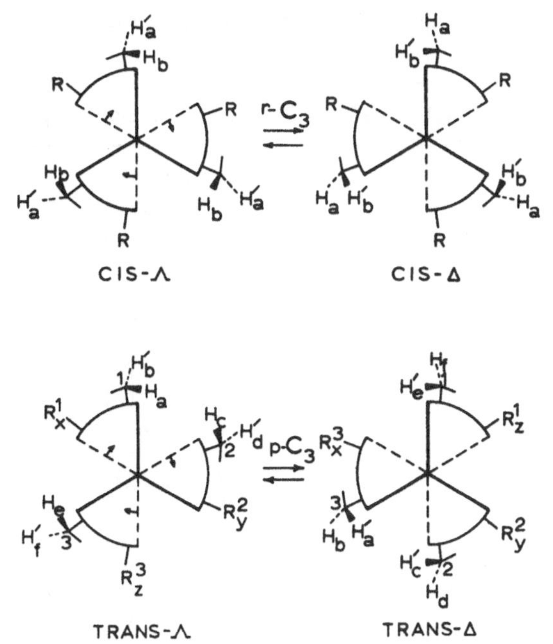

Fig. 7. Site rearrangement scheme of a M(A−B*)$_3$ type complex for the trigonal twist mechanism or averaging set A$_6$[36)]

cence of the four lines above −47 °C has been attributed to rapid S$_2$C−N bond rotation.[41)] Ligand exchange studies demonstrate that the exchange processes shown in Figs. 5 and 6 are intramolecular in origin.[19)]

Similar DNMR studies have been carried out on other M(dtc)$_3$ complexes where M = Ru(III),[37, 39)] Mn(III), V(III), Ga(III), In(III)[38)] and Co(III).[37)] All of these complexes are nonrigid due to metal-centered intramolecular rearrangement and kinetic parameters are given in Table 4. The kinetic parameters are all for enantio-merization, however, the mechanism or averaging set cannot be determined unambig-uously for every complex. If it is assumed that all M(dtc)$_3$ complexes of the same metal rearrange by the same mechanism, for example, Fe(CH$_3$, Bzdtc)$_3$ and Fe(Bz,-Bzdtc)$_3$, then the following mechanistic conclusions can be made: iron(III), (IV), ruthenium(III), vanadium(III) and manganese(III) rearrange by averaging set A$_6$; gallium(III) and indium(III) by A$_5$ or A$_6$; and cobalt(III) by A$_5$, A$_6$, A$_7$ or A$_8$. Complexes of chromium(III) and rhodium(III) are rigid up to 84 and 200 °C, respec-tively, so no mechanistic conclusions can be made. Only enantiomerization rates can be measured for complexes with symmetrical ligands via diastereotopic substi-tuents (see Section II).

It is argued[38)] that all of the nonrigid M(dtc)$_3$ complexes rearrange by averaging set A$_6$ in spite of the several ambiguous results noted above. This conclusion is based on the similarity of the $\Delta S^{\ddagger}$ values (all near zero) and on solid state results which shows a significant distortion toward trigonal prismatic geometry (the assumed transition state of the trigonal twist mechanism) for all M(dtc)$_3$ complexes (see Sec-tion V).

Table 4. Kinetic parameters for intramolecular metal-centered inversion for $M(dtc)_3$ complexes in CD_2Cl_2 solution

Complex[1]	$\Delta H^{\ddagger}$, kcal/mol	$\Delta S^{\ddagger}$, eu[4]	$\Delta G^{\ddagger}$, kcal/mol (temp, °C)	Ref.
$V(BzBzdtc)_3$	< 8.2		< 7.7 (−103)	38)
$V(MePhdtc)_3$	< 8.2		< 7.7 (−98)	38)
$Mn(BzBzdtc)_3$	11.0 ± 1.0	1.5 ± 5.0	10.6 ± 0.2 (−35)	38)
$Mn(MePhdtc)_3$	9.8 ± 1.0		9.1 ± 0.5 (−50)	38)
$Ga(BzBzdtc)_3$	< 8.6		< 8.1 (−85)	38)
$In(BzBzdtc)_3$	< 8.6		< 8.1 (−85)	38)
$Cr(EtEtdtc)_3$	> 17.1		> 16 (84)[2]	38)
$Fe(BzBzdtc)_3$	10.3 ± 1.0	4.1 ± 5.0	9.3 ± 0.2 (−54)	19)
$Fe(MePhdtc)_3$	8.7 ± 1.0	1.7 ± 5.0	8.5 ± 0.2 (−80)	19)
$Fe(pyrdtc)_3$	7.6 ± 1.7		7.1 ± 0.3 (−103)	19)
$Fe(BzBzdtc)_3^{\ddagger}$	8.4 ± 2.0		7.9 ± 0.4 (−88)	19)
$Co(BzBzdtc)_3$	25.5 ± 1.0	4.1 ± 5.0	23.6 ± 0.2 (168)[3]	19)
$Rh(BzBzdtc)_3$	> 27		> 25.3 (200)[3]	19)
$Ru(EtEtdtc)_3$	10.3	−8 ± 5	12.8 ± 0.2 (22)	39)
$Ru(BzBzdtc)_3$	11.6	−7 ± 5	13.8 ± 0.2 (42)[2]	39)
$Ru(MeBzdtc)_3$	11.1	5)	13.3 ± 0.2 (15)[2]	39)
$Ru(MePhdtc)_3$	10.1	5)	12.1 ± 0.2 (9)[2]	39)

1) Abbreviations given in Appendix.
2) $CDCl_3$ solution.
3) $NO_2C_6D_5$ solution.
4) Value assumed to be 3 eu where not given.
5) $\Delta S^{\ddagger}$ assumed to be −7.5 eu.

Many of the $M(dtc)_3$ complexes are paramagnetic and therefore have rather large isotropic NMR shifts.[25] This phenomenon usually enhances signal resolution (Figs. 5 and 6) over analogous diamagnetic complexes and has greatly simplified mechanistic analyses. Most of the studies reviewed hereafter involve diamagnetic complexes and the increased complexity of the analyses will be evident.

Several important results have come from the $M(dtc)_3$ studies in addition to the operation of the trigonal twist mechanism. First, the rate of enantiomerization is found to be greatly dependent on metal ion with the following order: In, Ga, V > Mn > Fe > Ru > Co > Rh with Cr at least greater than Ru. This trend including ligand effects will be considered in greater detail in Section VI. Secondly, $Ru(dtc)_3$ complexes represent the only class of stereochemically nonrigid ruthenium(III) tris chelates. All others which have been examined to date including $Ru(RT)_3$ and $Ru(R, R-\beta-dik)_3$ types are rigid on the NMR timescale (*vide infra*). Thirdly, $Co(dtc)_3$ complexes are the only class of tris chelates whose rate of enantiomerization has been determined both by DNMR and polarimetry. A recent study[44] reports the optical resolution and polarimetric kinetic results for $Co(pyrdtc)_3$ in $CHCl_3$ solution. The kinetic parameters are in good agreement with those determined by DNMR for $Co(BzBzdtc)_3$ and are $\Delta H^{\ddagger} = 23.4 \pm 2.1$ kcal/mol, $\Delta S^{\ddagger} = -0.8 \pm 7.0$ eu polarimetrically[44] and $\Delta H^{\ddagger} = 25.5 \pm 1.0$ kcal/mol, $\Delta S^{\ddagger} = 4.1 \pm 5.0$ eu by DNMR[19] for the above compounds, respectively.

B. Tris(tropolonates)

The α-substituted tropolonates *2* which are tris chelate complexes with a MO_6 coordination core have received considerable study by DNMR.[27, 45−46] The complexes are of the $M(A-B)_3$ and $M(A-B^*)_3$ types and therefore the averaging sets of Eaton *et al.*[26, 27] in Table 1 are applicable. The metal ions and α-R-substituents used in these studies include M = Al(III), Ga(III), Co(III), V(III), Mn(III), Ru(III), Rh(III) and Ge(IV); R = isopropyl (C_3H_7) and isopropenyl (C_3H_5), however, only complexes of Al(III), Ga(III), and Co(III) have yielded definitive mechanistic information.[27, 45−46] On the basis of line shape changes of the methyl resonances these complexes can be classed kinetically as follows: stereochemically nonrigid complexes which attain the fast-exchange limit of inversion and/or isomerization

- (i)   below *ca.* 0 °C [V(III), Mn(III), Ga(III)]
- (ii) below *ca.* 100 °C [Al(III), Co(III)], and
- (iii) above *ca.* 100 °C [Ge(IV)] and stereochemically rigid complexes [Rh(III), Ru(III)].[27, 45−46]

The DNMR spectra of the Al(III), Ga(III) and Co(III) complexes reveal two coalescence regions which are most clearly evident in the methyl DNMR spectra of

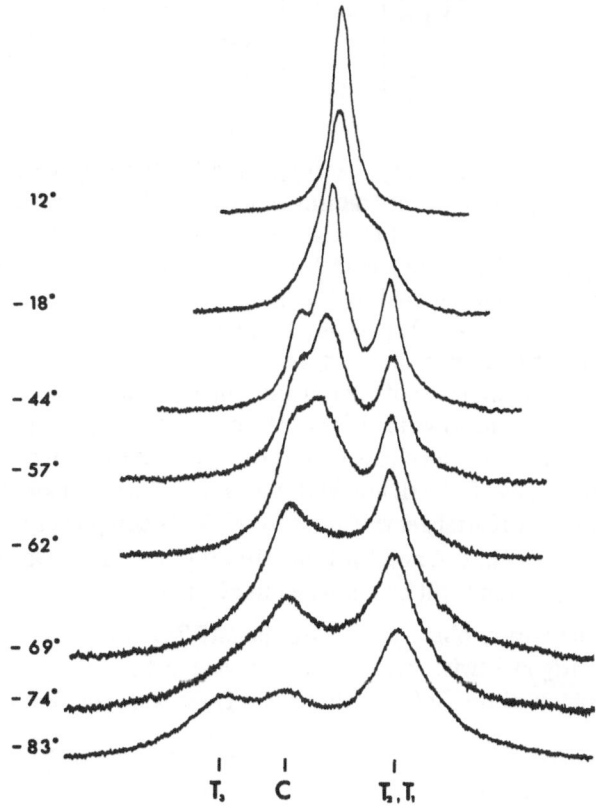

Fig. 8. Methyl spectra (100 MHz) of cis (C) and trans (T) $Ga(C_3H_5T)_3$ in dichloromethane solution. Chemical shifts of the centers of spin doublets are T_1, T_2, 3.64; C, 3.45; T_3, 3.35 ppm upfield of dichloromethane at $-83°C$. From Ref.[46]

$Ga(C_3H_5T)_3$ which is shown in Fig. 8.[46] The low temperature process (LTP) occurs between −83 and −44 °C and the (HTP) which results in coalescence of the three remaining lines occurs above −44 °C. Complexes of V(III) and Mn(III) do not show two regions of coalescence and their DNMR spectra indicate is-trans isomerization as does the (HTP) of the others. The Ge(IV) complexes show only a rather feature-less (LTP) which has general similarity to the (LTP) of the Co(III) and Al(III) complexes.[46] All of these processes have been shown to be intramolecular.

The observed averaging patterns of the (LTP) for $M(C_3H_5T)_3$ complexes of Al(III), Co(III), and Ga(III) are entirely consistent with either averaging set A_2 or A_6 (see Table 1).[i] In Fig. 8 the (yz) site permutation (after the notation in 6) of the T isomer is evident in the (LTP). The $C\Lambda \rightleftarrows C\Delta$ inversion is undetectable by NMR. The DNMR spectra of $Ga(C_3H_7T)_3$ show a pairwise averaging of eight methyl doublets to four in the (LTP) which is only consistent with averaging set A_6 which requires the site interchanges given in 11 for the T isomer (notation according to 7).[46] The C isomer shows the (rs) site permutation. Note that this averaging set is

$$
\begin{bmatrix}
1\text{-yr} \\
1\text{-ys} \\
2\text{-xr} \\
2\text{-xs} \\
3\text{-zr} \\
3\text{-zs}
\end{bmatrix}
\begin{array}{c} \longrightarrow \\ \longleftarrow \end{array}
\begin{bmatrix}
1\text{-zs} \\
1\text{-zr} \\
2\text{-xs} \\
2\text{-xr} \\
3\text{-ys} \\
3\text{-yr}
\end{bmatrix}
\qquad 11
$$
$$
\quad T\Lambda \qquad\qquad T\Delta
$$

the same as was found for $[Fe(CH_3, Bzdtc)_3]^+$ and the trigonal twist is the most satisfactory physical process or mechanism which gives this averaging set. The (LTP) DNMR spectra of $Al(C_3H_7T)_3$ and $Co(C_3H_7T)_3$ were also found to be consistent only with averaging set A_6 or the trigonal twist mechanism, however, with these complexes extensive computer analysis was necessary.[27] In addition, the rearrange-ment mechanism of $[Ge(C_3H_7T)_3]^+$ is provisionally assigned as a trigonal twist.[46] The kinetic parameters for inversion and/or isomerization are listed in Table 5.[46] Only the $M(C_3H_5T)_3$ complexes of Al(III) and Co(III) were subjected to total line-shape analysis, whereas the kinetic order of the other $M(RT)_3$ complexes is estimated from coalescence points. No mechanistic information was extracted from the non-rigid $M(RT)_3$ complexes of Mn(III) or V(III) or the rigid Ru(III) and Rh(III) complexes.[46] The order of rearrangement rates for $M(RT)_3$ complexes is for inversion: $Ga > Co \overset{\sim}{>} Al > Ge > Rh, Ru$; and for isomerization: $V \sim Mn \sim Ga > Co \overset{\sim}{>} Al > Ge > Rh, Ru$.[46] This trend will be discussed in Sections V and VI.

The most noteworthy observation with the $M(RT)_3$ complexes in addition to operation of the trigonal twist mechanism for several is the surprisingly low activation energies for inversion of the Co(III) complexes. These were the first nonrigid Co(III)

---

[i] The DNMR spectra of Al(III) and Co(III) require a computer fit for this conclusion whereas the Ga(III) spectra clearly show the site averaging pattern (Fig. 8).

Table 5. Kinetic parameters for Rearrangement of $M(RT)_3$ complexes[27]

Complex M, R	Solvent	Process	E_a kcal/mole	$\Delta S^{\ddagger}$	$\Delta G^{\ddagger}$ 298°
Al, C_3H_5	$C_2H_2Cl_4$	$C \to T$	16.9 ± 1.1	-4.3 ± 3.2	17.5 ± 1.4
	CH_2Cl_2	$T\Delta \to T\Lambda$	13.2 ± 1.4	-7.4 ± 4.8	14.9 ± 1.2
Al, C_3H_7	$C_2H_2Cl_4$	$C\Delta \to C\Lambda$	10.5 ± 3.6	-16 ± 12	14.6 ± 5.0
	$C_2H_2Cl_4$	$T\Delta \to T\Lambda$	12.3 ± 2.0	-11 ± 7	14.9 ± 2.8
	$C_2H_2Cl_4$	$C \to T$	25.8 ± 4.0	21 ± 12	18.7 ± 5.3
Co, C_3H_5	$CDCl_3$	$C \to T$	15.5 ± 1.1	-5.3 ± 3.7	16.5 ± 1.0
	$CDCl_3$	$T\Delta \to T\Lambda$	16.7 ± 0.9	5.4 ± 3.9	14.3 ± 0.7
Co, C_3H_7	$CHCl_3$	$C\Delta \to C\Lambda$	14.3 ± 1.0	-2.1 ± 3.5	14.3 ± 1.5
	$CHCl_3$	$T\Delta \to T\Lambda$	16.1 ± 1.3	4.8 ± 3.9	14.1 ± 1.8
	$CHCl_3$	$C \to T$	15.7 ± 1.4	-5.1 ± 4.2	16.6 ± 1.6

complexes discovered[46] and at 25 °C they enantiomerize $\sim 10^{11}$ faster than does $Co(CH_3, CH_3\text{-}\beta\text{-dik})_3$ and $\sim 10^7$ faster than does $Co(Bz, Bzdtc)_3$.[19] In general, the tris-tropolonates rearrange faster than the analogous tris-β-diketonates but slower than the analogous tris-dithiocarbamates with the notable exception of Co(III). This and similar trends will be discussed in Sections V and VI.

C. Tris(β-diketonates)

This class of complex given by structure 3 has been the most extensively studied tris chelate by both DNMR and classical techniques. Thusly a large body of kinetic data is available for rigid and nonrigid complexes, however, in not a single case has a unique mechanism been established, in spite of several elegant mechanistic studies on some Co(III) complexes.[12, 13] The difficulty in establishing a unique rearrangement mode is mostly due to the fact that bond rupture process seem to be operative. The various five coordinate transition states which result have similar energies and combinations of mechanisms apparently occur leading to complicated NMR averaging patterns or isomerization kinetics.[12, 13] The only rearrangement mode which has been uniquely established to date for any tris chelate is the trigonal twist as already discussed in Sections III.A and B for $M(dtc)_3$ and $M(RT)_3$ complexes, respectively.

The rigid or "slow" complexes will be considered first. Kinetic parameters for isomerization and/or racemization can be determined after separation of geometric or optical isomers usually by chromatographic techniques. Partial resolution of $M(CH_3, CH_3\text{-}\beta\text{-dik})_3$ complexes of Co(III), Rh(III), Cr(III),[13, 47] and Ru(III);[47] and complete resolution of Co(III) by repeated chromatography on D-lactose and careful crystallization[48] have been achieved. Kinetic parameters for intramolecular optical inversion of these resolved isomers in non-coordinating solvents are reported in Table 6. Geometric isomers of $M(CH_3, Ph\text{-}\beta\text{-dik})_3$ complexes where M = Co(III),[13, 49] Cr(III)[15, 49] and Rh(III)[49] have been separated and the former two subjected to kinetic analysis.[13, 15] Geometric isomers of $M(CH_3, CF_3\text{-}\beta\text{-dik})_3$

Table 6. Kinetic parameters for rearrangement of rigid $M(R_1 R_2\text{-}\beta\text{-dik})_3$ complexes

Complex $M(R_1 R_2)$	Solvent	Process	$\Delta H^{\ddagger}$ kcal/mol	$\Delta S^{\ddagger}$ eu	Ref.
$Co(CH_3CH_3)$	PhCl	Inversion	34.0 ± 0.6	14 ± 0.6	50)
$Cr(CH_3CH_3)$	PhCl	Inversion	34.2 ± 0.9	7 ± 2	50)
$Ru(CH_3CH_3)$	PhCl	Inversion	$> \sim 39.2^1)$	$-$	50)
$Rh(CH_3CH_3)$	PhCl	Inversion	$> \sim 42.3^1)$	$-$	50)
$Co(CH_3CF_3)$	$CHCl_3$	$C \rightarrow T$	30.0 ± 0.6	8.6 ± 1.7	15)
$Co(CH_3Ph)$	PhCl	$T \rightarrow C$	32.0 ± 0.6	8.8 ± 1.8	13)
	PhCl	$C \rightarrow T$	32.0 ± 0.6	11.0 ± 1.3	
	PhCl	$C\Delta \rightarrow C\Lambda$	32.2 ± 0.7	11.9 ± 2.0	
	PhCl	$T\Delta \rightarrow T\Lambda$	31.6 ± 0.9	10.0 ± 2.6	
$Co(CH_3iPr)$	PhCl	$T \rightarrow C$	31.9 ± 1.1	6.3 ± 2.9	12)
	PhCl	$C \rightarrow T$	32.2 ± 1.4	8.7 ± 3.9	
	PhCl	C (Racemization)	29.9 ± 0.6	1.5 ± 1.6	
	PhCl	T (Racemization)	31.6 ± 0.7	6.8 ± 1.8	

^{1}) Values are actually for E_a in kcal/mol.

complexes of Co(III) and Cr(III) have also been separated[13, 15] and the former kinetically investigated.[15] Partial resolution of cis-$Cr(CH_3, Ph\text{-}\beta\text{-dik})_3$,[15] cis- and trans-$Co(CH_3, Ph\text{-}\beta\text{-dik})_3$,[13, 50] and cis- and trans-$Co(CH_3, iPr\text{-}\beta\text{-dik})_3$[12] has also been achieved. Kinetic parameters for rearrangement are reported in Table 6.

It was recognized by Fay and Piper[15] in 1964 that mechanistic information can in principle be derived from a knowledge of the rates of isomerization and optical inversion for the same complex providing both reactions are due to the same mechanism. This sort of analysis was not applied, however, until 1970 when independently Gordon and Holm[12] and Girgis and Fay[13] separated and partially resolved the four geometric and optical isomers of $Co(CH_3, iPr\text{-}\beta\text{-dik})_3$ and $Co(CH_3, Ph\text{-}\beta\text{-dik})_3$, respectively. The kinetic parameters for isomerization and racemization (or inversion)[j] are shown in Table 6 for these complexes. These complexes isomerize and racemize at slightly different rates and activation parameters are closely similar for both processes which indicates that they proceed via the same mechanism(s). Six microscopic rate constants are required to kinetically describe the system comprised of the four isomers and are defined by

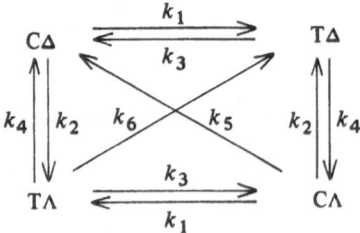

^{j}) The rate constant for optical inversion determined from a $\log[\alpha]$ vs time plot is twice that of the rate constant for direct $\Lambda \rightarrow \Delta$ inversion.[6]

and have been evaluated for both complexes.[12, 13] The ratios of the microscopic rate constants as well as ratios of isomerization rates to racemization rates are found to depend on the rearrangement mechanism.

Gordon and Holm[12] have thoroughly described the various rearrangement mechanisms for this type of system and have derived the appropriate rate ratios. The results of the two independent kinetic analyses are in good agreement and lead to the conclusion that the reaction mechanism is of the bond rupture variety with a high percentage of TBP-axial transition states.[12, 13]

Very recently[2] the above type of kinetic analysis has been applied to the rigid complex Ru(act)$_3$, 12, which has the chiral β-diketone ligand (+)-3-acetylcamphor. The four diastereomers (TΔ, TΛ, CΔ and CΛ) can readily be separated in optically

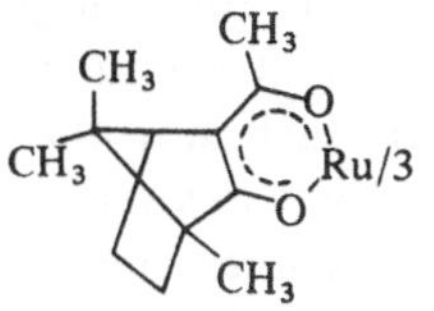

12

pure form and during kinetic experiments their presence and relative abundance can be determined by liquid chromatography.[2] In this case 12 microscopic rate constants are necessary to describe all possible interconversions among diastereomers. These were determined and the appropriate rate ratios were used to establish the most probable mechanism(s). For the CΔ and CΛ isomers the data show a reasonably good fit to bond rupture mechanisms with either SP-axial or 50% TBP-axial and 50% TBP-equatorial transition states. No single mechanism was found to fit the T isomer, however.[2]

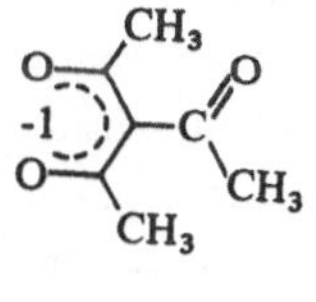

13

The above kinetic analyses of rigid complexes all indicate the operation of bond rupture processes. This result has also been found with the interesting β-diketone known as triac, 13, where the possibility of linkage isomerization exists.[51] Intramolecular bond rupture processes should result in similar rates for inversion and linkage isomerization in Co(triac)$_3$. Activation parameters for optical inversion in Co(triac)$_3$ are $\Delta H^{\ddagger} = 30.6 \pm 0.4$ kcal/mol and $\Delta S^{\ddagger} = 10 \pm 1$ eu whereas for linkage isomerization in Co(triac-d$_3$)$_3$ are 38.9 kcal/mol and 28 ± 4 eu, respectively.[51] The slightly higher activation enthalpy for linkage isomerization could result from internal ligand rotations which are necessary for isomerization. Also it is possible that ligand exchange could account for linkage isomerization although preliminary experiments with Co(CH$_3$, CH$_3$-β-dik)$_2$(triac-d$_3$) suggest otherwise.[52]

Much DNMR work has also been done on nonrigid tris(β-diketonates) but as with their rigid counterparts no unique rearrangement modes have been established and again bond rupture processes are usually found. The first DNMR experiments were carried out on $M(CH_3CF_3$-β-dik)$_3$ complexes where M = Al(III), Ga(III), and In(III) by Fay and Piper.[15] The ^{19}F DNMR spectra of the Ga(III) complex is shown in Fig. 1a. No mechanistic conclusions other than the elimination of the trigonal twist pathway as the sole rearrangement mode could be made, however, these experiments were the first of this type and are therefore of historic significance. The observed geometric isomerization in these complexes is thought to occur by bond rupture processes because the rate of the Al(III) complex was faster in more polar media.[15] Fay and Piper[15] pointed out the need for measuring both isomerization and racemization rates in order to gain mechanistic information. Jurado and Springer[22] showed that enantiomerization rates for nonrigid symmetrical tris chelates can conveniently be measured via diastereotopic site exchange. They observed the CH_3 DNMR of Al(iPr-β-dik)$_3$ (Fig. 2) which show the coalescence of diastereotopic CH_3 doublets as a result of rapid enantiomerization. A precise total line shape analysis yielded an Ea = 19 ± 1 kcal/mol which remained constant as the concentration was varied, showing that the kinetic process is first order and intramolecular.

The most detailed mechanistic analysis on nonrigid complexes which made use of kinetic data for isomerization and inversion was carried out by Hutchison et al.[53] on complexes of the type $M(Bz, iPr$-β-dik)$_3$ where M = Al(III) and Ga(III). The isopropyl CH_3 groups serve as probes for isomerization and inversion. For equilibrium mixtures of C and T isomers of the Al(III) and Ga(III) complexes in chlorobenzene the slow exchange NMR spectra (93 and 31 °C, respectively) reveal thirteen and nine of the sixteen possible methyl signals, respectively. DNMR spectra show a rather complicated coalescence pattern which for both complexes results in a single methyl spin doublet at the fast exchange limit (156 and 105 °C, respectively).[53] DNMR spectra for the Al(III) complex are shown in Fig. 9 along with several calculated spectra.[53] Mechanistic information could not be extracted directly from changes in signal multiplicity alone but rather required a complex line shape analysis which was carried out using site exchange matrices[k] for the various twist and bond rupture mechanisms and for random exchange. Several of the computer simulations are shown in Fig. 9[53] for the Al(III) complex. Careful analysis of the simulated line shapes led to the exclusion of the following mechanisms as sole reaction pathways for either complex: twists about the r-C_3 and p-C_3 axes only; bond rupture producing TBP-eq transition states; random exchange of ligand environments; for Ga(III), bond rupture via TBP-ax transition states. Satisfactory agreement was found for the following mechanisms: twists involving less than ca 50% and from ca. 25 to 75% rotation about the r-C_3 and p-C_3 axes of the Al(III) and Ga(III) complexes, respectively; bond rupture via SP-ax transition states; for Al(III), bond rupture producing TBP-ax transition states.[53] Kinetic results for the probable mechanisms are listed in Table 7.

---

[k] The Whitesides-Lisle EXCNMR computer program was employed; Whitesides, G. M. and Fleming, J. S., J. Amer. Chem. Soc. 89, 2855 (1967); Lisle, J. B., S. B. Thesis, M. I. T., 1968.

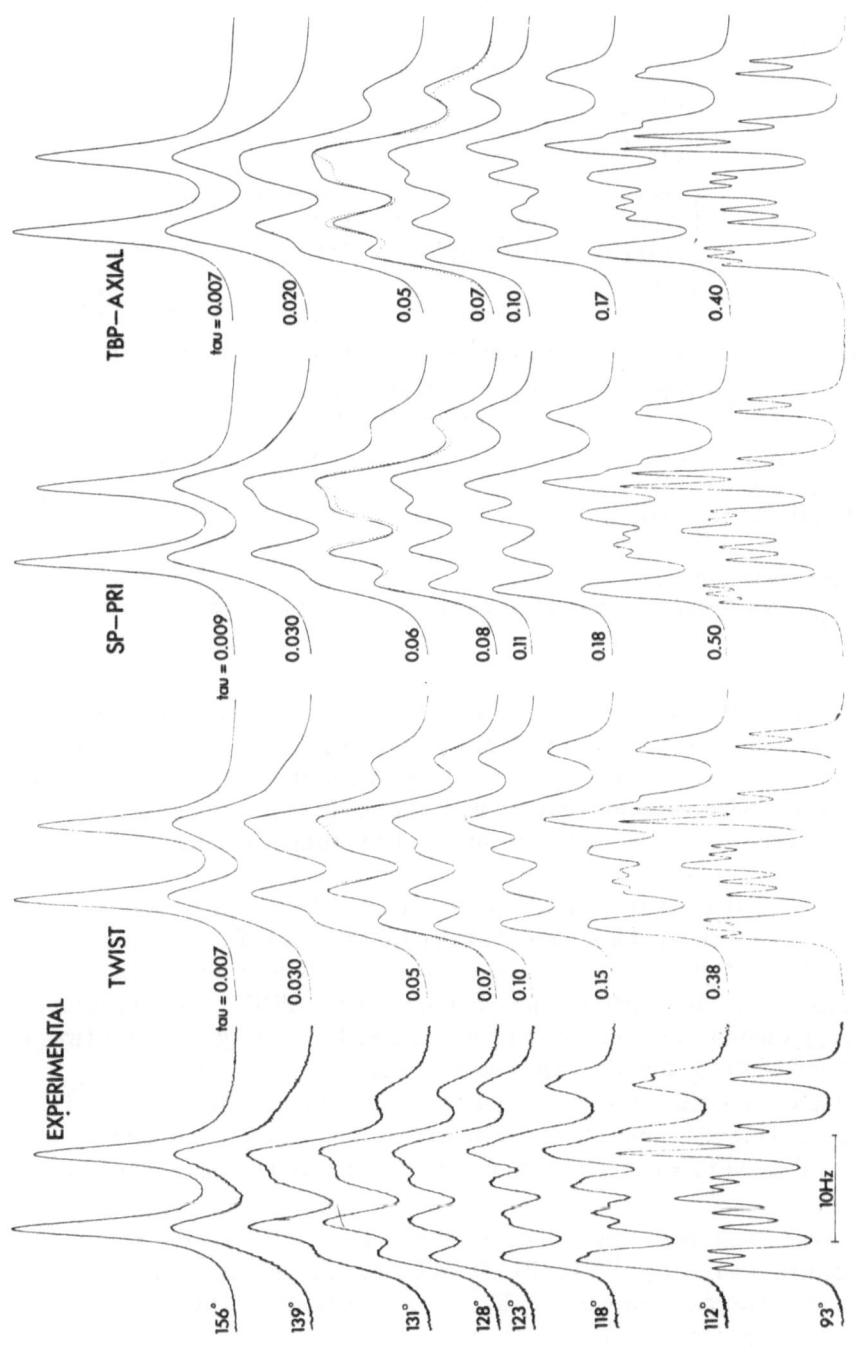

Fig. 9. Experimental methyl spectra Al(Bz, iPr-β-dik)₃ in chlorobenzene and simulated spectra for various rearrangement mechanisms: twist, 100% rotation about the imaginary C_3 axes of cis and trans isomers; SP–PRI, SP-primary process; experimental spectrum at 128°. τ values are in seconds. From Ref.[53]

Table 7. Kinetic parameters for rearrangement of nonrigid $M(R_1R_2\text{-}\beta\text{-dik})_3$ complexes and mixed β-diketonate complexes

Complex $M(R_1R_2)$	Solvent	Process	E_a kcal/mol	$\Delta S^{\ddagger}$ eu	Ref.
Al(CH$_3$CF$_3$)	CDCl$_3$	C → T	23.5 ± 1.8	–	15)
Ga(CH$_3$CF$_3$)	CDCl$_3$	C → T	20.8 ± 1.6	–	15)
In(CH$_3$CF$_3$)	CDCl$_3$	C → T	< 13.5	–	15)
Al(BziPr)	PhCl	twist [1])	27.6 ± 1.4	12.9 ± 3.5	53)
	PhCl	TBP-ax [1])	30.2 ± 1.4	19.5 ± 3.5	
	PhCl	SP-primary [1,2])	29.6 ± 1.4	17.6 ± 3.5	
Ga(BziPr)	PhCl	twist [1])	20.1 ± 1.2	2.9 ± 3.4	53)
	PhCl	SP-primary [1,2])	20.5 ± 1.2	3.6 ± 3.4	
Al(CH$_3$CH$_3$)$_2$(CF$_3$CF$_3$)	CH$_2$Cl$_2$	CH$_3$ site exchange	18.4 ± 0.7	10.0 ± 2.4	55)
Al(CH$_3$CH$_3$)(CF$_3$CF$_3$)$_2$	CH$_2$Cl$_2$	CF$_3$ site exchange	21.3 ± 0.7	10.7 ± 2.2	55)
	C$_6$H$_6$	CF$_3$ site exchange	19.0 ± 1.3	2.9 ± 4.2	
Al(CH$_3$CH$_3$)$_2$(PhPh)	o-C$_6$H$_4$Cl$_2$	CH$_3$ site exchange	22.0 ± 0.6	0.9 ± 1.5	55)
Ga(CH$_3$CH$_3$)$_2$(CF$_3$CF$_3$)	CH$_2$Cl$_2$	CH$_3$ site exchange	15.3 ± 1.2	3.9 ± 4.6	55)
Ga(CH$_3$CH$_3$)$_2$(PhPh)	C$_6$H$_6$	CH$_3$ site exchange	20.2 ± 3.2	2.2 ± 9.3	55)

[1]) Isomerization and inversion.
[2]) SP-primary denotes bond rupture via SP transition state by the simplest ligand motion [see Ref.[12)].

This entire analysis depends on correct signal assignment and chemical shift extrapolation, both of which could be in error. This difficulty could account for the mechanistic uncertainty, however, it is reasonably certain that at least some mechanisms can be excluded. The entire procedure is extremely time consuming and in light of the inconclusive results will probably not be profitable to apply to other systems.

Linkage isomerization rates were measured by DNMR for Al(triac)$_3$ and Ga(triac)$_3$ and estimated to be at least 1/40 and 1/800 times as fast as rearrangement rates, respectively. This result is interpreted to mean that bond rupture is unlikely for rearrangement of the Ga(III) complex but possible for Al(III).[53)] It is difficult to know exactly what rate differences are to be expected for rearrangement of M(triac)$_3$ and M(Bz, iPr-β-dik)$_3$ so the triac experiments are not conclusive, however, they strongly imply that Ga(Bz, iPr-β-dik)$_3$ rearranges by twist mechanisms whereas the analogous Al(III) complex most likely rearranges via bond rupture. It should be pointed out that multiple rearrangement mechanisms cannot be excluded by any of these results.

NMR experiments have also been carried out on M(CH$_3$, CF$_3$-β-dik)$_3$ complexes of Mn(III) and V(III).[54)] The CH$_3$ DNMR spectra of the Mn(III) complex shows the complete coalescence of four lines to one, indicating that C—T isomerization is occurring. The rate constant was estimated to be 8×10^2 sec^{-1} at 70 °C (the coalescence temperature) whereas the analogous Mn(CH$_3$, Ph-β-dik)$_3$ complex which has similar DNMR behavior isomerizes with a rate constant of ca. 1×10^3 sec^{-1} at 80 °C.[54)] The V(III) complex was found to be rigid up to 100 °C.[54)] No mechanistic information was obtained from these studies. Table 8 contains relative rearrangement rates of β-diketonate and α-R-tropolonate complexes all of which contain the MO_6 core. Although a detailed analysis of the relative rearrangement rates will be made in Sections V and VI, several important trends will be mentioned here. Rate data for MO_6 type complexes of Al(III), Ga(III), V(III), Mn(III), and Co(III) give kinetic series *14* for α-RT and $R_1 R_2$-β-dik type ligands.

$$\alpha\text{-RT} > (R_1 R_2 = CF_3, CH_3) > (CH_3, Ph) \sim (CH_3, CH_3) \sim (Bz, iPr) \sim$$
$$\sim (CH_3, iPr) \qquad\qquad 14$$

Table 8. Relative rearrangement rates of MO_6 complexes. From Ref.[46]

Ligand	Process	Rate order: $M[k,$[1]$] sec^{-1}$ (°C); E_a, kcal/mol]	Ref.
α-RT[2]	Inv	Ga [$\sim 10^2 (-65°)] >$ Co $[10^2, 17] \gtrsim$ Al $[80,13]$ > Ge > Si, Rh, Ru	27, 45, 46) [3]
	Isom	V $\sim$ Mn $\sim$ Ga > Co[4] $[4.0, 16] \gtrsim$ Al[4] $1.0, 17]$ > Ge > Si, Rh, Ru	27, 45, 46) [3]
CH_3CF_3-β-dik	Isom	Fe, In [> 36(−57°), < 14] > Mn [$\sim 10^3 (70°)]$ > Ga $[38(62°),21]$ > Al $[34(103°),24]$ > V > Co[4] $[5 \times 10^{-8}, 31]$ > Ru $[2 \times 10^{-12}, 33]$ > Rh $[< 10^{-8}$ (163°), > 42]	54, 15)
CH_3-Ph-β-dik	Isom	Mn $[8 \times 10^2$ (80°)] > V > Co[4] $[5 \times 10^{-9}, 33]$	54, 13)
BziPr-β-dik	Isom + inv	Sc > Ga[5] [$\sim 8,20]$ > Al[5] $[\sim 10^{-4}, \sim 29]$	53)
CH_3CH_3-β-dik	Inv	Co $[8 \times 10^{-10}, 35] \gtrsim$ Cr $[2 \times 10^{-11}, 35]$ > Ru $[< 3 \times 10^{-6}$ (135°), > 39] $\gtrsim$ Rh $[< 2 \times 10^{-6}$ (165°), > 42]	8)
mixed β-dik [6]	Site exchange	Ga $[8 \times 10^2, 15]$ > Al $[80, 18]$	55)
mixed β-dik [7]	Site exchange	Ga $[8 \times 10^{-2}, 20]$ > Al $[2 \times 10^{-3}, 22]$	55)

[1]) Measured or extrapolated value at 25 °C unless otherwise noted.
[2]) Data for R = C_3H_5 complexes.
[3]) Ito, T., Tanaka, N., Hanazaki, I., and Nakagura, S.: Inorg. Nucl. Chem. Lett. *5*, 781 (1969).
[4]) C → T; T → C kinetic parameters are similar.
[5]) Calculated using average values of the kinetic data obtained by line shape analysis.
[6]) $M(CH_3CH_3$-β-dik$)_2$ $(CF_3CF_3$-β-dik).
[7]) $M(CH_3CH_3$-β-dik$)_2$(PhPh-β-dik).
[8]) Fay, R. C., Girgis, A. Y., and Klabunde, U.: J. Amer. Chem. Soc. *92*, 7056 (1970).

For each type of ligand the rate dependence on metal ion is shown in Table 8 and the following observations can be made. For metal ions with a d^0 or d^{10} electronic configuration the rate of rearrangement increases with metal ionic radius as shown in series *15* in which ionic radii are in parentheses;

$$\text{In(III) (0.79), Sc(III) (0.73)} \gg \text{Ga(III) (0.62)} > \text{Al(III) (0.53)} \qquad 15$$

Rates for d^2–d^6 complexes show no trend with ionic radius, however, on descending a column in the periodic table the rates always decrease: Fe(III) > Ru(III) and Co(III) > Rh(III).

D. Mixed Ligand Complexes

Mixed ligand complexes are usually of the types $M(A-A)_2(B-B)$ and $M(A-B)_2(C-C)$ with and without diastereotopic probes. The permutational analysis and site interchanges have been worked out for the latter and are summarized in Tables 2 and 3. Most work has been done with nonrigid mixed β-diketonate complexes of Al(III) and Ga(III). Initially the $M(A-A)_2(B-B)$ complexes were examined by DNMR and the nonequivalent terminal A groups were found to coalesce. This site exchange was studied kinetically for the complexes listed in Table 7.[55] There has been much discussion in the literature concerning whether enantiomerization accompanies terminal A group exchange,[6, 56] however, careful analysis reveals that the relative rates of site exchange and enantiomerization depend on the mechanism.[56] Fortman and Sievers[56] have shown that twist motions about the p-C_3, all i-C_3 equally, and all axes equally give ratios of the rate of enantiomerization to the rate of terminal group exchange of 1:1, 3:2, and 4:3, respectively. Similarly, ratios of 3:2, 3:1, and 0:1 result for rupture via all TBP transition states equally, TBP-axial, and TBP-equatorial, respectively; and a ratio of 1:1 for bond rupture via all SP-axial transition states by the primary process.[56]

Careful kinetic analysis[57, 58] of CF_3 site exchange and i-propyl CH_3 exchange in $Al(CF_3, CF_3\text{-}\beta\text{-dik})_2(CH_3 CH_3\text{-}\beta\text{-dik})$ and $Al(CF_3, CF_3\text{-}\beta\text{-dik})_2(iPr, iPr\text{-}\beta\text{-dik})$, respectively, show very similar activation parameters for the two processes.[57, 58] This result although derived from two similar complexes strongly implies that bond rupture via SP-axial transition states or the p-C_3 twist mechanisms are operative.[6] This sort of analysis clearly shows how DNMR kinetic measurements and mechanistic considerations can lead to unique choices of rearrangement pathway.

A series of nonrigid $Fe(R_1 R_2 dtc)_2(S_2 C_2 Z_2)$ complexes, *16*, where $R_1 R_2 = CH_3$ CH_3; Et, Et; CH_3, Ph, and $(CH_2)_5$ and Z = CF_3, CN

16

have recently been synthesized and subjected to analysis by DNMR.[30, 32), 1)] As with the $M(R_1 R_2 dtc)_3$ complexes, two types of intramolecular processes can account for the stereochemical nonrigidity: ligand centered S_2C-N bond rotation and metal centered processes. The proton DNMR spectra of $Fe(Et, Etdtc)_2(S_2C_2(CN)_2)$ which are shown in Fig. 10[32), m)] reveal two distinct kinetic processes hereafter referred to

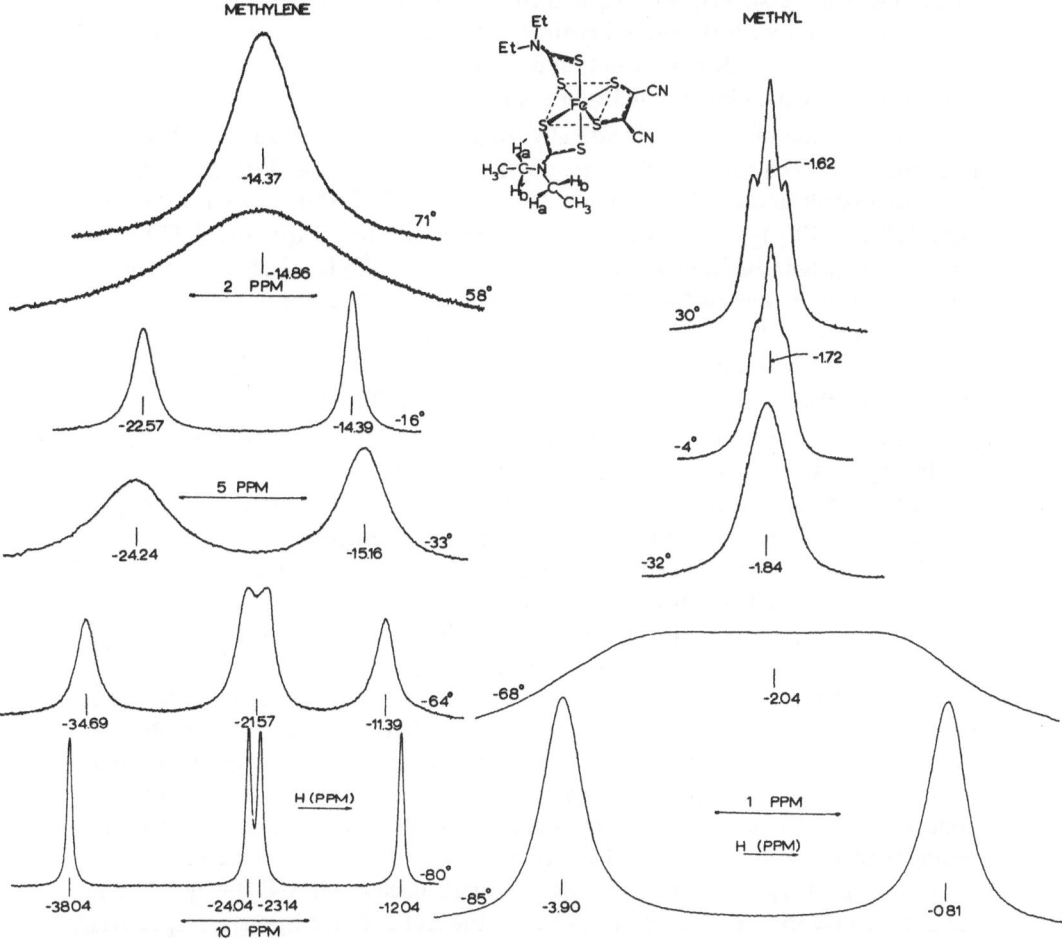

Fig. 10. Temperature dependence of the 100-MHz PMR spectrum of $Fe(Et, Etdtc)_2(S_2C_2(CN)_2)$ in CD_2Cl_2 solution. Sweep widths vary and are indicated below each series of spectra. Chemical shifts are downfield of TMS reference. From Ref.[32)]

---

1) These complexes are paramagnetic and posses a singlet-triplet spin state equilibrium in solid and solution phases[30, 32)]: Pignolet, L. H., Patterson, G. S., Weiher, J. F., Holm, R. H.: Inorg. Chem. *13*, 1263 (1974).

m) The DNMR spectra show excellent resolution of the four nonequivalent methylene environments. As mentioned earlier, paramagnetism in complexes usually results in large isotropic shifts which greatly simplify DNMR studies.[25)] If this complex was diamagnetic a complicated spin-spin coupled methylene pattern would probably result and subsequent analysis may be difficult.

as the (LTP) and (HTP). The (HTP) which is only observable for the methylene resonances has been assigned to S_2C-N bond rotation[32, 41] and will not be considered further here. The (LTP) averages two of the four nonequivalent methylene resonances H_a, H_b, H_a', H_b' and interchanges the two nonequivalent methyl environments. This coalescence pattern was originally explained by either S_2C-N bond rotation which averages H_a with H_a' and H_b with H_b' or metal-centered inversion which averages H_a with H_b and H_a' with H_b'. Both processes average the methyl environments. If S_2C-N bond rotation is operative in the (LTP), other metal centered processes would be required to account for the (HTP).

Consideration of an unsymmetrically substituted dtc ligand allows a distinction to be made between these mechanistic possibilities. In this case three geometric isomers are possible and are depicted in 6 The ^{19}F and ^1H DNMR spectra of Fe(CH$_3$, Phdtc)$_2$(S$_2$C$_2$(CF$_3$)$_2$) are shown in Fig. 11 and clearly reveal that the (LTP) does not result in geometric isomerization; therefore, the (LTP) is metal centered in origin. Resonances due to the cc isomer can be assigned because this isomer which lacks C_2 symmetry (see 6) shows spin-spin coupling between the nonequivalent CF$_3$ groups and requires equal intensity CH$_3$ resonances. The following stereochemical and site interchange information is evident from the (LTP) in Fig. 11.

(i) The ct $\rightleftarrows$ tc interconversion occurs as evidenced by the x $\leftrightarrow$ y peak coalescence but by a path which does not involve the cc isomer.

(ii) The rearrangement reaction of the cc isomer averages the CH$_3$ environments while simultaneously *not* averaging the CF$_3$ environments.

An additional reasonable constraint is that the same type of mechanism must apply to the processes in (i) and (ii). The observed configurational change (ct $\rightleftarrows$ tc) and site interchange pattern of the cc isomer, (mn) after 9, are consistent with only two averaging sets from the list in Table 2, A_2' and A_7'. Averaging set A_7' also requires enantiomerization which, of course, is NMR undetectable in this complex. The incorporation of a diastereotopic probe on the $S_2C_2Z_2$ ligand would allow a clear distinction between these two averaging sets as shown by the signal multiplicity changes listed in Table 3, however, this is synthetically difficult and has not been done. Note that the presence of a diastereotopic probe in the dtc ligand, a trivial synthetic task, will not permit a distinction between A_2' and A_7' (see Table 3).

Averaging set A_7' is given by the trigonal twist mechanism (p-C$_3$) which is depicted in Fig. 3 via transition states 1l and 5 for the Λct and Δcc isomers, respectively. Note that the configurational change Λcc $\rightleftarrows$ Δcc results in averaging environments a and b but not A and B. Averaging set A_2' is best considered to result from a mechanism which requires simultaneous rotation of each chelate ring 180° about its local C_2 axis passing through an idealized hexagonal planar transition state. The trigonal twist mechanism on the other hand involves a more plausible motion via an idealized trigonal prismatic transition state, a geometry which has been found experimentally for several MS$_6$ core complexes containing $S_2C_2Z_2$ type ligands.[59] Also the trigonal twist has been established for several M(dtc)$_3$ complexes (see Section III.A). These considerations provide strong support for averaging set A_7' and the trigonal twist mechanism. Kinetic parameters for complexes of type 16 are listed in Table 9. Note that the (HTP) has activation enthalpies which are 4–5 kcal/mole higher than for the (LTP).

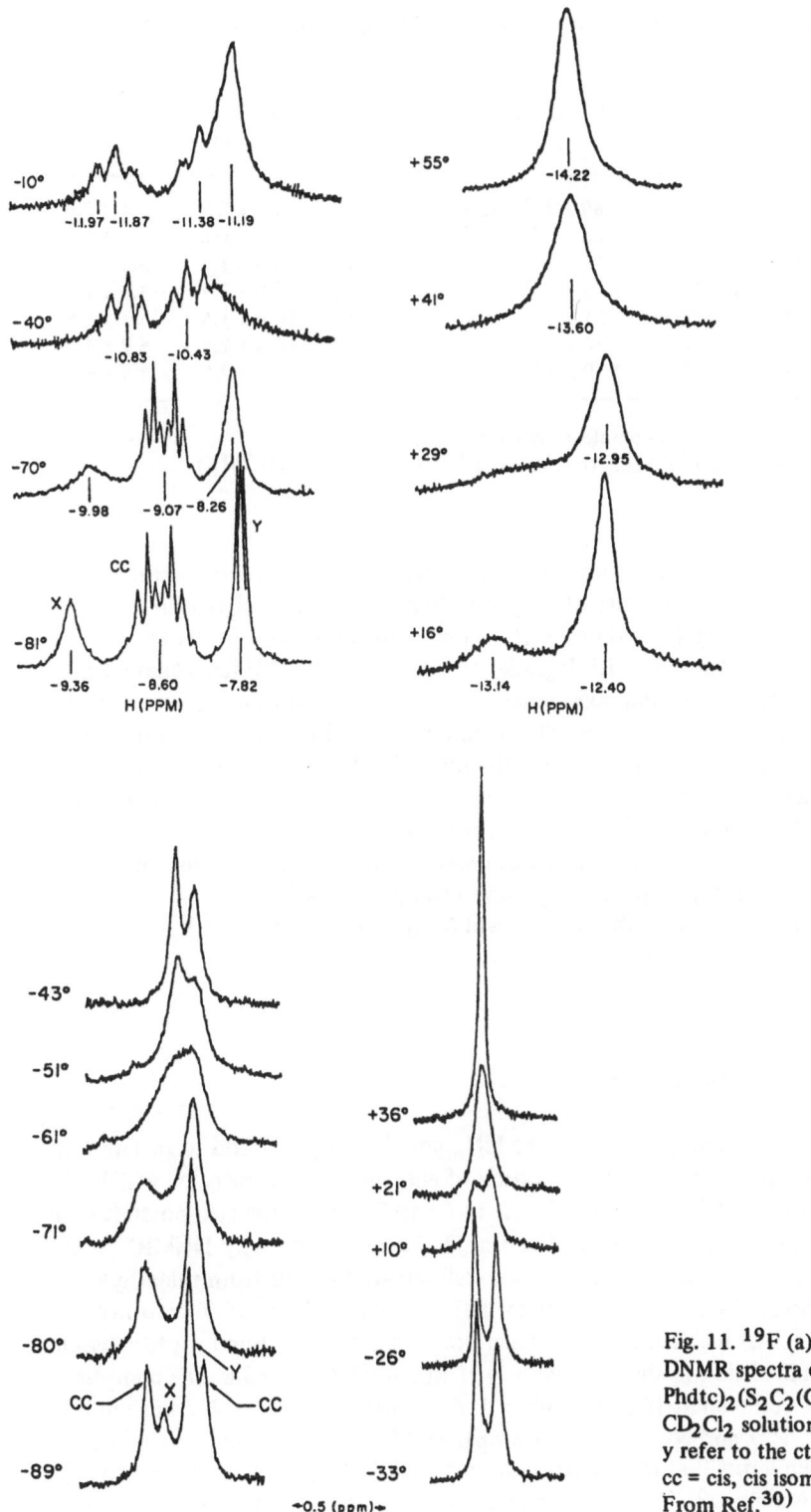

Fig. 11. ^{19}F (a) and ^{1}H (b) DNMR spectra of Fe(CH$_3$, Phdtc)$_2$(S$_2$C$_2$(CF$_3$)$_2$) in CD$_2$Cl$_2$ solution. Signals x, y refer to the ct, tc isomers; cc = cis, cis isomer, see 6. From Ref.[30])

Table 9. Kinetic data for rearrangement reactions of $Fe(R_1R_2dtc)_2(S_2C_2R_2)$ complexes in CD_2Cl_2 solution. From Ref.[32]

$R^{1)}$	R_1, R_2	Process	k^T, sec^{-1} (°)	$\Delta G^{\ddagger}{}_T,{}^{2)}$ kcal/mol	$\Delta H^{\ddagger}$, kcal/mol	$\Delta S^{\ddagger}$, eu
CF_3	CH_3, CH_3	LTP	1.89×10^2 (−50)	10.6 ± 0.2	9.2 ± 0.9	-6.1 ± 3.9
CF_3	C_2H_5, C_2H_5	LTP	7.85×10^2 (−50)	10.0 ± 0.2	8.3 ± 0.6	-7.5 ± 3.7
CN	C_2H_5, C_2H_5	LTP	3.11×10^3 (−50)	9.4 ± 0.2	8.6 ± 1.5	-3.4 ± 5.0
CF_3	CH_3, Ph	HTP	1.29×10^2 (25)	14.9 ± 0.2	14.0 ± 2.1	-3.0 ± 7.8
CF_3	C_2H_5, C_2H_5	HTP	4.11×10^1 (25)	15.2 ± 0.2	16.4 ± 1.5	4.1 ± 4.5
CF_3	$(CH_2)_5$	HTP	4.38×10^1 (25)	15.2 ± 0.2	16.6 ± 2.4	4.7 ± 7.5
CN	C_2H_5, C_2H_5	HTP	2.30×10^1 (25)	14.1 ± 0.2	12.5 ± 2.0	-5.5 ± 6.0

[1] Kinetic parameters for CF_3 complexes from Ref.[30].
[2] Calculated from k^T at T °C using the relation $k^T = (k_BT/h)exp(-\Delta G^{\ddagger}/RT)$.

Several additional studies have been reported with mixed ligand complexes. La Mar[60] has measured the rate of collapse of nonequivalent β-dik methyl signals in $Co(II)(CH_3, CH_3\text{-}\beta\text{-dik})_2(4,7\text{-dimethyl},1,10\text{-phenanthroline})$ and found an $E_a \approx 13$ kcal/mol and $\log A \approx 16$. Pignolet and co-workers[19, 61] have examined $Fe(II)(R_1R_2dtc)_2L$ type complexes where L = 1,10-phenanthroline, phen, or bis(1,2-diphenylphosphino)ethane, diphos. These complexes are high ($S = 2$) and low ($S = 0$) spin, respectively, and have activation enthalpies of < 7 and $> \sim 20$ kcal/mole, respectively. Fortman and Seivers[5, 62, 63] have reported preliminary DNMR results on the mixed β-diketonate $Al(CF_3, CF_3\text{-}\beta\text{-dik})(CH_3, CH_3\text{-}\beta\text{-dik})(t, C_4H_9, t\text{-}C_4H_9\text{-}\beta\text{-dik})$. No firm mechanistic conclusions have been reached, however, rate constants for the three types of terminal group exchange have been measured.[5]

Some of the rate data of this section will be analyzed in Sections V and VI.

E. Miscellaneous Tris(bidentates)

A number of catonic complexes with the MN_6 core have been found to racemize or possess DNMR properties. The racemization of $Ni(ethylenediamine)_3{}^{+2}$, which is actually the reaction $\Lambda(\lambda\lambda\lambda) \rightleftarrows \Delta(\delta\delta\delta)$ where λ and δ refer to the two possible conformations of the five-membered chelate ring, has been observed by DNMR[64]. At room temperature in aqueous solution two well resolved signals from the ethylenic protons are observed which coalesce at ca. 101 °C giving $\Delta G^{\ddagger} \approx 15.7$ kcal/mole. Intramolecular twist mechanisms have been postulated because bond rupture would cause nitrogen inversion in the analogous N,N'-dimethylethylenediamine complex which was not observed at 101°.[65] Other similar Ni(II) complexes, Ni(meso-butyl-enediamino)$_3{}^{+2}$ [66] and Ni(1,3-diaminopropane)$_3{}^{+2}$,[67] have been examined by DNMR with the result that racemization occurs slightly faster than for the ethyl-enediamino complex with $\Delta G^{\ddagger}$ at 50 and 72 °C equal to ca. 14.0 and 14.0 kcal/mol, respectively. A twist mechanism was again proposed for the former.[66] M(ethyl-enediamino)$_3^{+3}$ complexes of Cr(III), Co(III), and Os(III) are extremely inert to racemization.[1]

The tris-(phenanthroline), phen, and -(bipyridino), bipy, complexes of Fe(II) have been found to racemize predominately by an intramolecular pathway.[68, 69] The racemization also proceeds partly via ligand dissociation but at much slower rates.[68, 69] The mechanism for intramolecular racemization is thought to be a non-bond-breaking twist process because the planar, rigid phen ligand is not able to act as a monodentate ligand, however, the large $\Delta S^{\ddagger}$ value of +21 eu[68] is not consistent with the near zero or slightly negative values found for other "twisting" complexes (see Tables 4, 5 and 9). The low spin ($S = 0$) Fe(phen)$_3^{+2}$ complex has a $\Delta H^{\ddagger}$ of ca. 29 kcal/mol.[68, 69]

Numerous kinetic studies of racemization reactions have been performed with tris(oxalato) metal complexes, 17, M(OX)$_3^{-3}$ in aqueous solution where

$$17$$

M = Cr(III),[70] Co(III),[70] and Rh(III).[71] These reactions are complicated by acid and metal ion catalysis and in the case of Co(III) decomposition is significant.[70] In addition the OX^{-2} oxygen atoms are found to exchange with ^{18}O labeled water sometimes at rates comparable with racemization.[72]

The racemization of Cr(OX)$_3^{-3}$ and Co(OX)$_3^{-3}$ is found to be intramolecular because labeled OX^{-2} does not exchange with the complexed OX^{-2} in the time required for racemization.[71, 73] In neutral aqueous solution racemization of Cr(OX)$_3^{-2}$ occurs but the OX^{-2} oxygens do not exchange with ^{18}OH$_2$ and a twist mechanism has been proposed. In the presence of acid both racemization and oxygen exchange with ^{18}OH$_2$ are accelerated and therefore an acid catalyzed bond rupture process probably occurs.[72] Racemization of Co(OX)$_3^{-3}$ in neutral and acid solution presumably also occurs by a twist mechanism.[70] This result comes from extensive oxygen exchange studies involving both inner and outer OX^{-2} oxygen exchange rates.[70] In any case, both of these complexes are thought to racemize by a twist mechanism in neutral solution, in addition, definite acid catalysis occurs which at least in the case of Cr(OX)$_3^{-3}$ probably results from bond rupture. Activation parameters for optical inversion in neutral solution are listed in Table 10.

The only other tris(oxalato) complex to receive much attention is Rh(OX)$_3^{-3}$.[71,73] This complex undergoes racemization in acid solution which is not due to inter-molecular OX^{-2} exchange.[74] Oxygen exchange studies with ^{18}OH$_2$ show that the rate of racemization is similar to the rate of inner oxygen exchange and therefore an intramolecular bond rupture mechanism is favored[71] in contrast to the Co(III) complex, where racemization is much faster than inner oxygen exchange.

Table 10. Kinetic data for optical inversion of $M(OX)_3{}^{-3}$ complexes in neutral aqueous solution

Complex	$k \times 10^5$ sec^{-1}, ($t\,°C$)	E_a kcal/mole	$\Delta S^{\ddagger}$ [1] eu
$Cr(OX)_3{}^{-3}$ [2]	1.75, (18.2)	15.75	−23.7
$Co(OX)_3{}^{-3}$ [3]	1.88, (24.87)	27.1	6.7

[1] Ref. [6].
[2] Bushra, E., Johnson, C. H.: J. Chem. Soc. 1937 (1939).
[3] Ref. [70].

V. Influence of Solid State Geometry on Dynamics

The previous section has dealt with the kinetics and probable mechanisms for metal centered rearrangement reactions of tris chelate complexes. A general rationale of these results is best obtained by considering the β-diketonates, α-R-tropolonates and dithiocarbamates separately. These three classes have been the most extensively studied and have yielded the most firm mechanistic conclusions. It is obvious from the previous sections that rates and mechanisms depend markedly on metal ion and ligand type. We will approach the problem of explaining these results by first considering the effect of ground state geometry as determined from X-ray data on the kinetic parameters and mechanisms for rearrangement.

X-ray results are available for numerous tris chelate complexes, however, we will be primarily concerned with the three classes mentioned above. The most striking feature of the structural results is the adherence to $\sim D_3$ symmetry even in complexes which are severely distorted from the octahedral or trigonal antiprismatic D_{3d} [n] limit. Muetterties and Guggenberger[75] have recently pointed out that with the exception of about six tris(dithiolate) complexes, *18*, that are close to the D_{3h} (trigonal prismatic) limit, all structurally established tris chelates have D_3 or near

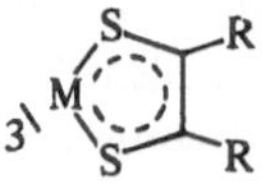

R = H, CN, Ph
18

D_3 symmetry and they are at or near the D_{3d} limit or along the $D_{3d} \rightleftarrows D_{3h}$ reaction path.[75] This observation has obvious importance with respect to the trigonal twist mechanism.

---

[n] D_{3d} symmetry will be used to describe trigonal antiprismatic geometry even though the tris chelate complexes have lower real symmetry due to the presence of the chelate rings.

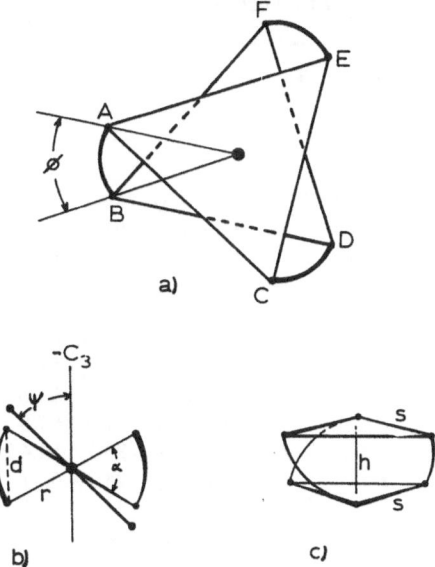

Fig. 12. Definition of solid state structural parameters for tris chelate complexes with $\sim D_3$ symmetry: ϕ is the twist angle and is the projection of the bite angle α onto the plane which is perpendicular to the C_3 symmetry axis, ψ is the pitch angle and is the angle subtended by the plane of the chelate ring and the C_3 symmetry axis; r, d, s and h are the metal ligand distance, bite distance, triangle edge, and the distance between the triangles, respectively. From Ref.[82]

In order to describe the geometry of tris chelate complexes the following shape parameters have been selected and are defined in Fig. 12. The angle α is the chelate bite angle and is related to the metal ligand distance, r, and the bite distance, d, by $d = 2r \sin(\alpha/2)$; the angle ϕ is the so called twist angle and is the projection of α on the plane which is perpendicular to the C_3 symmetry axis (see below); the angle ψ is the chelate ring or propeller pitch angle; and the distances s and h are the edge length of the triangle and intertriangle separation, respectively. The polar angle θ is also convenient in relating the various parameters and is the angle subtended by the M−L bond and the C_3 symmetry axis. All of these parameters have been used by others,[76−81] however, the importance of the pitch angle ψ in describing distortions away from idealized D_{3d} symmetry has not been emphasized. [38, 82] The following equations most conveniently relate the angles and distances:

$$\cos(\alpha/2) = \sin\theta \, \cos(\phi/2) \qquad (1)$$

$$\sin(\alpha/2) = \cos\theta/\cos\psi \qquad (2)$$

$$s = \sqrt{3} \, (r \sin\theta) \qquad (3)$$

$$\cos\psi = h/d \qquad (4)$$

Table 11. Structural parameters [1]) for tris chelate complexes of $\sim D_3$ symmetry

Complex	$\bar{r}$ Å	$\bar{d}$ Å	$\bar{s}$ Å	$\bar{h}$ Å	α deg.	ϕ deg.	$\bar{\psi}$ deg.	$\bar{s}/\bar{h}$	Footnote
(R₁R₂-β- diketonates(3)									
$Mg(CF_3CF_3)$	2.058	2.771	2.94	2.31	84.7	53.1	33.4	1.27	[2])
$Al(CH_3CH_3)$	1.892	2.718	2.64	2.24	91.8	60.9	34.7	1.18	[3])
$Sc(CH_3CH_3)$	2.070	2.715	2.96	2.34	82.0	47.3	30.3	1.26	[4])
$V(CH_3CH_3)\ \alpha$	1.979	2.749	2.79	2.30	88.0	55.5	33.0	1.21	[5])
$V(CH_3CH_3)\ \beta$	1.987	2.745	2.81	2.30	87.4	55.2	33.2	1.22	[5])
$Cr(CH_3CH_3)$	1.953	2.786	2.74	2.29	91.0	60.3	34.8	1.20	[6])
$Cr(CH_3CH_3)$	1.956	2.793	2.76	2.27	91.1	61.5	35.7	1.22	[7])
$Fe(CH_3CH_3)$	1.992	2.744	2.86	2.32	87.1	53.6	32.1	1.23	[8])
$Co(CH_3CH_3)$	1.897	2.849	2.62	2.29	97.3	67.9	36.4	1.14	[3])
$Ru(CH_3CH_3)$	1.992	2.907	2.79	2.34	93.7	64.4	36.2	1.19	[9])
$Rh(CH_3CH_3)$	1.992	2.946	2.77	2.37	95.3	66.1	36.3	1.17	[10])
$Cr(atc)_3$	1.969	2.831	2.78	2.28	92.0	62.5	36.1	1.22	[11])
Averages					**90.1**	**59.0**	**34.4**	**1.21**	
Tropolonates (2, R = H)									
$Al(T)_3$	1.88	2.490	2.70	2.13	82.6	48.5	31.0	1.27	[12])
$Sc(T)_3$	2.102	2.523	3.04	2.30	73.8	34.0	24.1	1.32	[13])
$Fe(T)_3$	2.008	2.522	2.87	2.27	77.8	38.4	25.6	1.26	[14])
Averages					**78.1**	**40.3**	**26.9**	**1.28**	
(R₁R₂)dithiocarbamates (1)									
$Mn(EtEt)$	2.450	2.908	3.64	2.52	72.8	40.2	29.9	1.44	[15])
$Fe(EtEt)$	2.357	2.845	3.44	2.54	74.2	37.6	26.8	1.35	[16])
$Fe(EtEt),\ 79\ °K$	2.307	2.836	3.36	2.50	75.9	40.4	28.2	1.34	[16])
$Fe[(CH_2)_4]$	2.407	2.911	3.50	2.60	74.4	37.4	26.5	1.35	[17])
$Fe(CH_3Ph)$	2.312	2.819	3.38	2.47	75.1	40.7	28.9	1.37	[17])
$Co(EtEt)$	2.258	2.786	3.32	2.39	76.2	43.7	30.7	1.39	[18])
$Ru(EtEt)$	2.376	2.827	3.50	2.50	73.0	38.1	27.8	1.40	[19])
$In[(CH_2)_5]$	2.588	2.947	3.84	2.66	69.4	32.8	25.3	1.44	[20])
$Tl(CH_3CH_3)$	2.659	2.979	3.99	2.65	68.2	33.2	26.5	1.50	[21])
Averages					**73.2**	**38.2**	**27.8**	**1.40**	
Miscellaneous									
$V[S_2C_2(Ph)_2]_3$	2.338	3.059	3.07	3.05	81.7	1.9	1.2	1.01	[22])
$Mo[Se_2C_2(CF_3)_2]_3$	2.491	3.315	3.22	3.31	83.4	0	0	0.97	[23])
$Re[S_2C_2(Ph)_2]_3$	2.325	3.032	3.05	3.03	81.4	3.4	2.0	1.01	[24])
$Cr(OX)_3^{-2}$	1.965	2.591	2.83	2.18	82.5	50.5	32.8	1.30	[25])
$Co(OX)_3^{-2}$	1.922	2.580	2.77	2.13	84.3	54.1	34.3	1.30	[26])
$Rh(OX)_3^{-2}$	2.016	2.678	2.94	2.18	83.0	54.5	35.5	1.35	[27])
$Fe(phen)_3^{+2}$	1.969	2.814	3.12	2.31	91.4	60.7	34.8	1.35	[28])
$Ni(phen)_3^{+2}$	2.090	2.669	3.43	2.25	79.4	48.2	32.7	1.52	[29])
$Ni(en)_3^{+2}$	2.124	2.789	3.07	2.35	82.0	50.2	32.6	1.31	[30])

[1]) Some of the angular parameters (ϕ and ψ) may vary slightly from values reported in the literature due to the details of the calculation, especially the location of the $\sim C_3$ axis (see text).

[2]) Truter, M. R., Vickery, B. L.: J. Chem. Soc. Dalton, *1972*, 395.

[3]) Hon, P. K., Pfluger, C. E.: J. Coord. Chem. *3*, 67 (1973).

[4]) Anderson, T. J., Neuman, M. A., Melson, G. A.: Inorg. Chem. *12*, 927 (1973).

[5]) Morosin, B., Montgomery, H.: Acta Cryst. *B25*, 1354 (1969).

[6]) Wright, W. B., Meyers, E. A.: Cryst. Struct. Comm. *2*, 477 (1973).

[7]) Morosin, B.: Acta Cryst. *19*, 131 (1965).

[8]) Iball, J., Morgan, C. H.: Acta Cryst. *23*, 239 (1967).

[9]) Chou, G. K.-J., Sime, R. L., Sime, R. J.: Acta Cryst. *B29*, 2845 (1973).

[10]) Morrow, J. C., Parker, E. B., Jr.: Acta Cryst. *B29*, 1145 (1973).

[11]) Horrocks, W. D., Jr., Johnston, D. L., MacInnes, D.: J. Amer. Chem. Soc. *92*, 7620 (1970).

[12]) Muetterties, E. L., Guggenberger, L. J.: J. Amer. Chem. Soc. *94*, 8046 (1972).

[13]) Anderson, T. J., Neuman, G. A., Melson, G. A.: Inorg. Chem. *13*, 158 (1974).

[14]) Hamor, T. A., Watkin, D. J.: Chem. Commun. *1969*, 440.

[15]) Healy, P. C., White, A. H.: J. Chem. Soc. Dalton *1972*, 1883.

[16]) Leipoldt, J. G., Coppens, P.: Inorg. Chem. *12*, 2269 (1973).

[17]) Ref.[79]).

[18]) Merlino, S.: Acta Cryst. *B24*, 1441 (1968).

[19]) Ref.[82]).

[20]) Hauser, P. J., Bordner, J., Schreiner, A. F.: Inorg. Chem. *12*, 1347 (1973).

[21]) Abrahamson, H., Britton, D., Pignolet, L. H., in preparation.

[22]) Eisenberg, R., Gray, H. B.: Inorg. Chem. *6*, 1844 (1967).

[23]) Pierpont, C. G., Eisenberg, R.: J. Chem. Soc. *A 1971*, 2285.

[24]) Eisenberg, R., Ibers, J. A.: Inorg. Chem. *5*, 411 (1966).

[25]) van Niekerk, J. N., Schoening, F. R. L.: Acta Cryst. *5*, 499 (1952).

[26]) Butler, K. R., Snow, M. R.: J. Chem. Soc. *A, 1971*, 565.

[27]) Dalzell, B. C., Eriks, K.: J. Amer. Chem. Soc. *93*, 4298 (1971).

[28]) Zalkin, A., Templeton, D. H., Veki, T.: Inorg. Chem. *12*, 1641 (1973).

[29]) Frenz, B. A., Ibers, J. A.: Inorg. Chem. *11*, 1109 (1972).

[30]) Ul-Haque, M., Caughlan, C. N., Emerson, K.: Inorg. Chem. *9*, 2421 (1970).

$$h = 2r \cos \theta \qquad (5)$$

$$\theta = \arctan[2/\sqrt{3}\,(s/h)] \qquad (6)$$

Parameters were calculated from the reported positional coordinates and the crystallographic unit cell constants and are listed in Table 11 for selected complexes. For complexes which lacked crystallographic C_3 symmetry the $\sim C_3$ axis was defined as follows. The midpoints of the lines connecting the ligating atoms of each bidentate chelate were calculated and the plane containing these three midpoints constructed. The $\sim C_3$ axis is defined as the normal to this plane which contains the metal ion. The parameters ϑ, ψ and ϕ were then calculated.[o])

For octahedral geometry some of the structural parameters are fixed by definition and are $\alpha = 90°$, $\phi = 60°$, $\psi = 35.3°$, and $s/h = 1.22$. The ratio s/h is the so called compression ratio[78]) and is useful in describing the relative flatness of a polyhedron. For trigonal prismatic D_{3h} geometry $\phi = 0°$ and $\psi = 0°$. D_{3h} symmetry has no further requirements and the value of s/h depends on α. The only tris chelates which possess D_{3h} geometry are several neutral tris(dithiolenes) (Table 11) which

---

[o]) During the preparation of this article the author discovered by private communication from R. H. Holm that a similar but more extensive analysis has been carried out by M. R. Snow which will eventually be published in Coord. Chem. Rev.

have $\alpha \simeq 82°$ and $s/h \simeq 1.0$. It is not obvious, however, that $s/h = 1.0$ is a necessary condition for D_{3h} coordination or that compressed or flattened complexes ($s/h >$ 1.0) which have D_3 ground state geometries will not enantiomerize via D_{3h} transition states.[p] Indeed, the trigonal twist mechanism is most likely operative for highly compressed complexes [$M(dtc)_3$ and $M(RT)_3$, *vide infra*]. Trigonal antiprismatic or D_{3d} geometry requires $\phi = 60°$.

Most of the tris bidentate chelate complexes listed in Table 11 do not have octahedral geometry because the chelate bite angle α is usually less than 90°. This is contrary to the O_h symmetry commonly assumed for six coordinate complexes. Since most tris chelate complexes[q] and all of those listed in Table 11 possess $\sim D_3$ symmetry, the twist angle ϕ has most often been used to describe the extent of twist from trigonal antiprismatic D_{3d} ($\phi = 60°$) to trigonal prismatic D_{3h} ($\phi = 0°$) geometry. The parameter ϕ has accordingly been used to describe the likeliness of inversion via the trigonal twist mechanism since highly twisted complexes ($\phi \ll 60°$) are already near the presumed D_{3h} transition state geometry.[7, 8] Consideration of the twist angle ϕ alone, however, can lead to erroneous mechanistic predictions. Idealized O_h geometry has $\phi = 60°$, $\psi = 35.3°$ and $\alpha = 90°$ but also requires mutual orthogonality of the three chelate rings. If α is less than 90° it is possible to preserve the orthogonality by simply decreasing ϕ. In this case, however, ψ remains at 35.3°. For example, $\alpha = 75°$ [a value often found for $M(dtc)_3$ complexes] gives $\phi = 48°$ and $\psi = 35.3°$ for mutually orthogonal chelate rings. Indeed, α and ϕ can decrease to nearly 0° (hypothetically) while still retaining the orthogonality condition. This observation shows that the trigonal twist mechanism is best envisioned as the concerted rotation of the three chelate rings about their respective C_2 axes such that the average value of ψ goes from its ground state value through 0° (D_{3h} transition state) to the negative of its ground state value. The usual ϕ criterion for inversion via a trigonal twist can be misleading because a small ground state value of ϕ does not necessarily indicate a twist toward D_{3h} since α may be very small whereas a small value of ψ ($< 35.5°$) always does regardless of the size of α. Of course the ϕ criterion is valid and equivalent to the ψ criterion if the value of α is also considered. Indeed Eqs. (1) and (2) relate α, ϕ and ψ.

Examination of the results in Table 11 shows that the $M(\beta\text{-dik})_3$ complexes are all near the octahedral limit. The average value of α, ϕ, ψ and s/h is 90.1°, 59.0°, 34.4°, and 1.21, respectively, for the twelve complexes listed. The $M(T)_3$ complexes have averages of 78.1°, 40.3°, 26.9° and 1.30, respectively, for the same parameters of the three complexes listed which indicates a significant twist toward D_{3h} geometry as well as a slight flattening or compression of the coordination core. The $M(dtc)_3$ complexes have averages of 73.2°, 38.2°, 27.8°, and 1.40, respectively, for the nine complexes listed which also shows a twist toward D_{3h} and much greater flattening.

---

p) Stiefel and Brown[78] have suggested that highly compressed tris chelate complexes may enantiomerize via a hexagonal planar transition state because they have $s/h > 1.0$ when twisted to D_{3h} while maintaining the chelate ring dimensions.

q) Several tris chelate complexes show an additional tetragonal distortion due to the Jahn-Teller effect, *e.q.*, with Mn(III), and therefore do not have $\sim D_3$ symmetry. These distortions are not related to intramolecular reaction pathways and these complexes are omitted from this discussion.

The values of ψ are most important here because α varies from $\sim 90°$ to $\sim 70°$ for $M(\beta\text{-dik})_3$ and $M(dtc)_3$ complexes, respectively. Hence the amount of twist toward D_{3h} by the ϕ criterion is greatest for the $M(dtc)_3$ complexes whereas the ψ criterion shows that the $M(T)_3$ complexes are actually twisted slightly more. The smaller average ϕ value of the $M(dtc)_3$ complexes results because of the smaller average value of α [73.2 compared with 78.1 for $M(T)_3$]. However, since only three $M(T)_3$ complexes have been structurally characterized, the average values of α, ϕ and ψ are not statistically meaningful and overall comparisons with the $M(dtc)_3$ complexes are not valid.[r]

The degree of twist noted above in the ground state geometry of tris chelate complexes has been calculated by Kepert.[77] The calculation was based on the summation of ligand-ligand repulsion energies between the six donor atoms of the ML_6 core.[77] The calculation was carried out for a fixed value of α as a function of $\phi/2$. The minimum repulsion energy for values of $\alpha < 90°$ occurred for values of $\phi < 60°$ indicating that twist distortions toward D_{3h} geometry result because of small bite angles. A value of $90°$ for α had an energy minimum for O_h geometry. Kepert[77] did not report values of ψ, however, they are easily calculated from α and ϕ [Eqs. (1) and (2)] and indeed small chelate bite angles yield minimum repulsion energies for values of $\psi < 35.3°$. In fact, the calculations quite amazingly predict the experimentally observed twist and pitch angles for a number of complexes with the exception of the tris(dithiolenes) which have $\sim D_{3h}$ geometry. For the $M(dtc)_3$ series listed in Table 11 note that the smaller values of α correspond with the smaller values of ϕ and ψ.

The mechanisms for metal centered rearrangement should in principle be related to ground state geometry and it is found that the complexes which rearrange via the trigonal twist mechanism are just the ones which are distorted toward D_{3h} geometry. It should be noted, however, that these distortions are in no case larger than $\sim 32\%$ by the ψ criterion. Kepert's calculation also predicts that small bite ligands should have lower activation energies for inversion via a trigonal twist.[77] Hence, the tris(β-diketonates) which have $\sim O_h$ geometry rearrange by bond rupture pathways with the possible exception of $Ga(Bz, iPr-\beta\text{-dik})_3$ where a twist pathway is likely; whereas the tris(tropolonates) and tris(dithiocarbamates) are distorted toward D_{3h} geometry and rearrange via the trigonal twist pathway. At this time it can be concluded that the small bite angles of the five and four membered T and dtc chelate rings, respectively, are of primary importance in causing the operation of the trigonal twist mechanism.

It is less fruitful to use structural data to explain the relative rearrangement rates even for complexes which rearrange by the same mechanism. The most extensive series of constant mechanism complexes is of the $M(dtc)_3$ type and structural results are available for many of the complexes. A comparison of $\Delta H^{\ddagger}$ for enantiomerization to α, ϕ and ψ is made in Table 12. Although some of the $\Delta H^{\ddagger}$ values are within experimental error of each other, an overal relation exists. The higher $\Delta H^{\ddagger}$ values in general correspond to the less twisted (toward D_{3h}) complexes, however, the magnitude of the differences in angular parameters is hardly sufficient to account

---

[r] Preliminary structural results from $Co(T)_3$ indicate $\phi \sim 55°$, Eisenberg, R., cited in Ref.[27]

Table 12. Crystallographic parameters of the MS_6 core and kinetic parameters for metal-centered inversion of several $M(dtc)_3$ complexes

Complex	$\Delta H^{\ddagger}$ kcal/mol (± 1)	Angles[2] of MS_6 Core, deg. α	ϕ^{2}	ψ	Ref. of structure determination
In[$(CH_2)_5$dtc]$_3$	$\ll 8.6$[1]	69.4	32.8	25.3	[4]
Fe[$(CH_2)_4$dtc]$_3$	7.6[3]	74.4	37.4	26.5	[5]
Fe(MePhdtc)$_3$	8.7	75.1	40.7	28.9	[5]
Ru(EtEtdtc)$_3$	10.3	73.0	38.1	27.8	[6]
Mn(EtEtdtc)$_3$	11.0[1]	72.8	40.2	29.9	[7]
Co(EtEtdtc)$_3$	25.5[1]	76.2	43.7	30.7	[8]

[1] $\Delta H^{\ddagger}$ determined for analogous BzBzdtc complex Ref.[38].

[2] Average values calculated from the crystallographic coordinates given in the appropriate reference.

[3] Values differ slightly from those reported elsewhere due to the method of calculation, however, these values are correct for relative comparison.

[4] Hauser, P. J., Bordner, J., Schreiner, A. F.: Inorg. Chem. *12*, 1347 (1973).

[5] Ref. [79].

[6] Ref. [39].

[7] Healy, P. C., White, A. H.: J. Chem. Soc. Dalton 1883 (1972).

[8] Merlino, S.: Acta. Cryst. *B 24*, 1441 (1968).

for the much greater $\Delta \dot{H}^{\ddagger}$ value found for Co(III). Electronic effects must be invoked in this case and in general are important in determining relative rearrangement.[1] The next section will be concerned with d electronic effects.

Eaton *et al.*[46] have made a rather extensive comparison of the rearrangement rates and mechanisms for tris(β-diketonates) and tris (tropolonates). The following two structural points may be important in favoring the trigonal twist for the latter and bond rupture for the former:[46]

(i) The rigid, planar nature of the tropolonate ligand should tend to supress a bond rupture mechanism since more energy, in the form of M–O or M–O–C bond deformations, would be required to remove one end of the ligand from bonding distance to the metal than in the case of the internally flexible β-diketonates.

(ii) The relatively short bite distance of the tropolonate ligand (*ca.* 2.5Å) leads to polyhedron radius to polyhedron edge ratios which are close to the ideal value of 0.76 given by Day and Hoard[83] for D_{3h} geometry. The following ratios have been calculated assuming bite and M–O distances are the same in the D_{3h} as in the ground-state trigonal antiprismatic configuration: CoT_3, 0.74;[r] Co(acac)$_3$, 0.66; AlT_3, 0.76; Al(acac)$_3$, 0.69; GaT_3, 0.78 (estimated; MnT_3, ~ 0.78.[84, s]

The ground state geometry of the neutral tris(dithiolenes) is essentially D_{3h} (Table 11). This result is inconsistent with Kepert's[77] calculation since α is *ca.* 82°

[s] MnT_3 does not possess trigonal symmetry and the quoted distance ratio as well as the average twist angle (~ 49°) are not as meaningful as for strictly trigonal chelates.

and therefore the complexes should have geometries which are similar to the tris(tropolonates). The trigonal prismatic geometry of these complexes has been rationalized by interligand S–S attractions between the S atoms of each trigonal face.[59] However, the $M(dtc)_3$ complexes which also have an MS_6 core do not have D_{3h} coordination but rather have geometries exactly predictable by Kepert's calculation.

The remaining entries in Table 11 have parameters which are less easily interpreted in terms of rearrangement dynamics. Again the complexes with smaller values of α are the most twisted toward D_{3h}. The $M(OX)_3^{-3}$ complexes, however, show that values of α less than $90°$ while usually leading to values of ϕ less than $60°$ do not necessarily show twists toward D_{3h} by the pitch angle criterion.

VI. Influence of Electronic Configuration on Dynamics

A consideration of the d electronic configuration should in principle permit certain rate predictions for rearrangement reactions of tris chelate complexes. Electronic effects have been applied with some success in explaining the relative rates of ligand substitution reactions.[1, 85] For example, Basolo and Pearson[1] used a ligand field stabilization energy, $LFSE$, argumental such that relative activation energies for dissociative ligand substitution of six coordinate complexes are related to the differences in $LFSE$ between octahedral and trigonal bipyramidal or square pyramidal geometry. This procedure nicely accounted for the inertness of Co(III) and Cr(III) complexes. Taube[85] used electrostatic arguments based on d orbital occupation to account for the inertness of low spin Co(III) and Cr(III) toward associative ligand substitution. Both of these approaches require a knowledge of the reaction mechanism, *i.e.*, a specific transition state or activated complex geometry must be assumed. Likewise for intramolecular rearrangement reactions the mechanism must be known before any electronic effect arguments can be applied.

In the previous section it was noted that the trigonally twisted nature of the ground state geometry of $M(dtc)_3$ and $M(T)_3$ complexes was probably important in causing the operation of the trigonal twist mechanism. The tris(β-diketonates) are not distorted in this way and rearrange via bond rupture processes. Since several transition state geometries are possible and probable for rearrangements of the latter it is impossible to apply electronic configuration arguments with any degree of success. However, such arguments should be useful in accounting for the relative rearrangement rates of $M(dtc)_3$ and $M(T)_3$ complexes for which the trigonal twist mechanism via a trigonal prismatic transition state is extremely likely. Since the $M(dtc)_3$ series is the most extensive it will be used exclusively in the following discussion.

Recent ligand field calculations[76, 80] have been carried out on trigonal antiprismatic, TAP, trigonal prismatic, TP, and intermediate geometries which yield good estimates of the $LFSE$ for TAP and TP geometries for the various d electronic configurations. A complex which enantiomerizes by the trigonal twist mechanism can be assumed to proceed from TAP or near TAP geometry through a TP transition state and therefore the activation energy should be related to $LFSE(\text{TAP}) - LFSE(\text{TP})$ or

$\Delta LFSE$. This procedure has been successfully applied in establishing stereochemical preferences in a series of clathro-chelate complexes.[76, 81] The results have been applied to enantiomerization rates of $M(dtc)_3$ complexes and are shown in Table 13.[19, 38] The $\Delta LFSE$ values used here were determined from the calculation of

Table 13. $\Delta LFSE^{1)}$ values and $\Delta H^{\ddagger}$ for enantiomerization of dithiocarbamate complexes in CD_2Cl_2 Solution

Metal Ion, spin	No. of d electrons	$\Delta LFSE$ Dg	$\Delta H^{\ddagger}$ kcal/mol	Ref.
Fe(II), 2	6	0	$< 7.5^{2)}$	19)
Ga(III), 0	10	0	< 8.6	38)
In(III), 0	10	0	< 8.6	38)
V(III), 1	2	3.4	< 8.2	38)
Mn(III), 2	4	3.4	9.8–11.0	38)
Fe(III), 5/2	5	0	7.6–10.3 $^{3)}$	19)
Fe(III), 1/2	5	10.2		
Ru(III), 1/2	5	10.2	10.1–11.6	39)
Fe(IV), 1	4	6.8	8.4	19)
Cr(III), 3/2	3	6.8	> 16	38)
Fe(II), 0	6	13.6	$> 20^{4)}$	61)
Co(III), 0	6	13.6	25.5	19)
Rh(III), 0	6	13.6	> 27	19)

$^{1)}$ LFSE = TAP(LFSE) − TP(LFSE) after Gillum et al.[80]
$^{2)}$ $Fe(dtc)_2$phen.
$^{3)}$ $Fe(III)(dtc)_3$ complexes have a $S = 1/2 \rightleftarrows S = 5/2$ spin state equilibrium.
$^{4)}$ $Fe(dtc)_2(CO)_2$ in toluene-d_8.

Gillum et al.,[80] however, the more sophisticated calculation of Larsen et al.[76] yields very similar results.

The data reported in Table 13 do show a relation between $\Delta H^{\ddagger}$ and $\Delta LFSE$ which is in general agreement with the above assumption. For Mn(III) the $\Delta H^{\ddagger}$ values are slightly high which may result from the observed tetragonal distortion in the solid state which places these complexes slightly off the D_{3d}–D_{3h} reaction pathway. Since $\Delta LFSE$ is in units of D_q the complexes from the second row of the periodic table should enantiomerse slower than their first row counterparts. Hence the trends in $\Delta H^{\ddagger}$

$$(Ru(III) > Fe(III), Rh(III) > Co(III))$$

are expected and consistent with results from $M(CH_3, CH_3\text{-}\beta\text{-dik})_3$ complexes where for $\Delta H^{\ddagger}$

$$Rh(III) > Co(III) \text{ and } Ru(III) > Fe(III),$$

and for $M(RT)_3$ complexes where $Rh(III) > Co(III)$. The Fe(III) complexes possess a low spin $\rightleftarrows$ high spin equilibrium which complicates the $\Delta LFSE$ argument since $\Delta LFSE$ is 10.2 Dq and 0 D_q, respectively, for these spin states. The spin equilibrium is rapidly attained, however, which could mask any rate effects due to the equilibrium constant which is greatly dependent on the N-substituent.[19] In any case it is observed that the complexes which are predominately high spin enantiomerize faster than ones which are mainly low spin. This trend is also consistent with solid state data which shows that the predominately high spin complexes are twisted toward D_{3h} to a greater extent than the low spin complexes. In Table 11 note the structural parameters for $Fe(EtEtdtc)_3$ at $79°$ and $297°K$ where the complex is predominately low and high spin, respectively.

The *LFSE* analysis assumes that other factors such as bond energies, nonbonded interactions, solvent effects and geometries have essentially equal changes during enantiomerization for one metal to another. In effect, these obviously important factors must cancel so the effect of $\Delta LFSE$ dominates. Further the spin states are assumed to remain unchanged from TAP to TP geometry so electron pairing effects are eliminated. The fact that the *LFSE* analysis seems to qualitatively work is surprising but no more so than for ligand substitution reactions.

It is difficult to extend the *LFSE* conclusions to complexes of different ligand type since the ligand structural and electronic characteristics may result in large rate effects. For example, $Co(RT)_3$ complexes isomerize much faster than $Co(dtc)_3$ complexes in spite of the fact that all other $M(dtc)_3$ complexes rearrange faster than the analogous $M(RT)_3$ complexes. This observation has not been satisfactorily explained and possibly could involve low lying electronic excited states in $Co(RT)_3$.

Acknowledgements. Preprints of papers from R. H. Holm and R. L. Martin are gratefully acknowledged. The author also acknowledges many helpful discussions with Professors R. H. Holm and D. Britton which were invaluable to the preparation of parts of this paper. The assistance of Mr. H. Abrahamson in compiling the structural parameters and the dedicated work of Drs. L. Que, Jr., D. J. Duffy, and B. L. Edgar and Mr. M. C. Palazzotto has been greatly appreciated. Financial support from the National Science Foundation Grant GP-37795, the donors of the Petroleum Research Fund, administered by the American Chemical Society, the Research Corporation, and the University of Minnesota Graduate School is also gratefully acknowledged.

VII. References

[1] Basolo, F., Pearson, R. G.: Mechanisms of Inorganic Reactions, 2nd ed. New York: J. Wiley & Sons 1967.

[2] Everett, G. W., Jr., Horn, R. R.: J. Amer. Chem. Soc. *96*, 2087 (1974).

[3] Holm, R. H.: Accts. Chem. Res. *2*, 307 (1969).

[4] Binsch, G.: Top. Stereochem. *3*, 97 (1968).

[5] Fortman, J. J., Sievers, R. E.: Coord. Chem. Rev. *6*, 331 (1971).

[6] Serpone, N., Bickley, D. G.: Progr. Inorg. Chem. *17*, 391 (1972).

[7] Pignolet, L. H., La Mar, G. N., in: NMR of paramagnetic molecules (ed. G. N. La Mar, W. D. Horrocks, Jr., R. H. Holm), Chap. 8. New York: Academic Press 1973.

[8] Holm, R. H. in: Dynamic nuclear magnetic resonance spectroscopy (ed. L. M. Jackman, F. A. Cotton), Chap. 9. New York: Academic Press 1974.

[9] Holm, R. H., O'Connor, M. J.: Progr. Inorg. Chem. *14*, 241 (1971).

[10] Holm, R. H., Hawkins, C. J., in: NMR of Paramagnetic molecules (ed. G. N. La Mar, W. D. Horrocks, Jr., R. H. Holm), Chap. 7. New York: Academic Press 1973.

11) Muetterties, E., L., in: Reaction mechanisms in inorganic chemistry, MTP international review of science, (ed. M. L. Tobe), Vol. 9, Chap. 2. London: Butterworths 1972.

12) Gordon, II, J. G., Holm, R. H.: J. Amer. Chem. Soc. *92*, 5319 (1972).

13) Girgis, A. Y., Fay, R. C.: J. Amer. Chem. Soc. *92*, 7061 (1972).

14) Musher, J. I.: Inorg. Chem. *11*, 2335 (1972).

15) Fay, R. C., Piper, T. S.: Inorg. Chem. *3*, 348 (1964).

16) For example, DNMR3: Binsch, G., Kleier, D. A., Dept. of Chemistry, University of Notre Dame, Notre Dame, Indiana 46556 USA; or the Whitesides-Lisle EXCNMR program: Whitesides, G. M., Fleming, J. S.: J. Amer. Chem. Soc. *89*, 2855 (1967); Lisle, J. B.: S. B. Thesis, M. I. T., 1968.

17) Meakin, P., Muetterties, E. L., Jesson, J. P.: J. Amer. Chem. Soc. *94*, 5271 (1972).

18) Meakin, P., Muetterties, E. L., Jesson, J. P.: J. Amer. Chem. Soc. *95*, 75 (1973).

19) Palazzotto, M. C., Duffy, D. J., Edgar, B. L., Que, L., Jr., Pignolet, L. H.: J. Amer. Chem. Soc. *95*, 4537 (1973).

20) Bailar, J. C.: J. Inorg. Nucl. Chem. *8*, 165 (1958).

21) Mislow, K., Rabin, M.: Top. Stereochem. *1*, 1 (1965).

22) Jurado, B., Springer, C. S., Jr.: Chem. Commun., 85 (1971).

23) Springer, C. S., Jr.: private communication reported in Ref.[6].

24) Palazzotto, M. C., Pignolet, L. H.: Inorg. Chem. *13*, 1781 (1974).

25) Jesson, J. P., in: NMR of paramagnetic molecules (ed. G. N. La Mar, W. D. Horrocks, Jr., R. H. Holm), Chap. 1. New York: Academic Press 1973.

26) Eaton, S. S., Eaton, G. R.: J. Amer. Chem. Soc. *95*, 1825 (1973).

27) Eaton, S. S., Hutchison, J. R., Holm, R. H., Muetterties, E. L.: J. Amer. Chem. Soc. *94*, 6411 (1972).

28) Springer, C. S., Jr., Sievers, R. E.: Inorg. Chem. *6*, 852 (1967).

29) Råy, P., Dutt, N. K.: J. Indian Chem. Soc. *20*, 81 (1943).

30) Pignolet, L. H., Lewis, R. A., Holm, R. H.: J. Amer. Chem. Soc. *93*, 360 (1971).

31) Muetterties, E. L.: J. Amer. Chem. Soc. *92*, 7369 (1970); Springer, C. S., Jr.: J. Amer. Chem. Soc. *95*, 1459 (1973) and references cited therein.

32) Pignolet, L. H., Lewis, R. A., Holm, R. H.: Inorg. Chem. *11*, 99 (1972).

33) Musher, J. I.: J. Amer. Chem. Soc. *94*, 5662 (1972).

34) Longuet-Higgins, H. C.: Mol. Phys. *6*, 445 (1963).

35) Palazzotto, M. C., Pignolet, L. H.: J. C. S. Chem. Commun., *1972*, 6.

36) Duffy, D. J., Pignolet, L. H.: Inorg. Chem. *11*, 2843 (1972).

37) Pignolet, L. H., Duffy, D. J., Que, L., Jr.: J. Amer. Chem. Soc. *95*, 295 (1973).

38) Que, L., Jr., Pignolet, L. H.: Inorg. Chem. *13*, 351 (1974).

39) Duffy, D. J., Pignolet, L. H.: Inorg. Chem. *13*, 2045 (1974).

40) Golding, R. M., Tennant, W. C., Bailey, J. P. M., Hudson, A.: J. Chem. Phys. *48*, 764 (1968).

41) Edgar, B. L., Duffy, D. J., Palazzotto, M. C., Pignolet, L. H.: J. Amer. Chem. Soc. *95*, 1125 (1973).

42) Pasek, E. A., Straub, D. K.: Inorg. Chem. *11*, 259 (1972).

43) Cauguis, G., Lachenal, D.: Inorg. Nucl. Chem. Letters *9*, 1095 (1973).

44) Gahan, L. R., Hughes, J. G., O'Connor, M. J.: J. Amer. Chem. Soc. *96*, 2271 (1974).

45) Muetterties, E. L., Alegranti, C. W.: J. Amer. Chem. Soc. *91*, 4420 (1969).

46) Eaton, S. S., Eaton, G. R., Holm, R. H., Muetterties, E. L.: J. Amer. Chem. Soc. *95*, 1116 (1973).

47) Collman, J. P., Blair, R. P., Marshall, R. L., Slade, L.: Inorg. Chem. *2*, 576 (1963).

48) Sun, J. Y.: Dissertation Abstr. *28B*, 4482 (1968).

49) Fay, R. C., Piper, T. S.: J. Amer. Chem. Soc. *84*, 2303 (1962).

50) Fay, R. C., Girgis, A. Y., Klabunde, U.: J. Amer. Chem. Soc. *92*, 7056 (1970).

51) Fay, R. C.: personal communication reported in Ref.[6].

52) Klabunde, U., Fay, R. C.: Abstracts of Papers No. INOR-37, 156th National Meeting of the American Chemical Society, Atlantic City, N. J., September 1968.

53) Hutchison, J. R., Gordon, II, J. G., Holm, R. H.: Inorg. Chem. *10*, 1004 (1971).

54) Gordon, II, J. G., O'Connor, M. J., Holm, R. H.: Inorg. Chim. Acta *5*, 381 (1971).

55) Case, D. A., Pinnavaia, T. J.: Inorg. Chem. *10*, 482 (1971).
56) Fortman, J. J., Sievers, R. E.: Inorg. Chem. *6*, 2022 (1967).
57) Springer, C. S., Jr.: personal communication reported in Ref.[6].
58) Ruggiero, L. A. W., Jurado, B., Springer, C. S., Jr.: Abstracts of Papers No. INOR-159, 161st National Meeting of the American Chemical Society, Los Angeles, Calif., March 1971.
59) Eisenberg, R.: Progr. Inorg. Chem. *12*, 295 (1970).
60) La Mar, G. N.: J. Amer. Chem. Soc. *92*, 1806 (1970).
61) Edgar, B. L.: Ph. D. dissertation, University of Minnesota, Minneapolis, Minn., 1974.
62) Fortman, J. J., Sievers, R. E.: Abstracts of Papers No. M57, 155th National Meeting of the American Chemical Society, San Francisco, Calif., April 1968.
63) Fortman, J. J.: personal communication reported in Ref.[6].
64) Ho, F. F.-L., Reilley, C. N.: Anal. Chem. *42*, 600 (1970).
65) Ho, F. F.-L., Reilley, C. N.: Anal. Chem. *41*, 1835 (1969).
66) Evilia, R. F., Young, D. G., Reilley, C. N.: Inorg. Chem. *10*, 433 (1971).
67) Sarneski, J. E., Reilley, C. N.: Inorg. Chem. *13*, 977 (1970).
68) Basolo, F., Hayes, J. C., Neumann, H. M.: J. Amer. Chem. Soc. *76*, 3807 (1954).
69) Dowley, P., Garbett, K., Gillard, R. D.: Inorg. Chim. Acta *1*, 278 (1967).
70) Broomhead, J. A., Lauder, I., Nimmo, P.: J. Chem. Soc. *A*, 645 (1971) and references cited therein.
71) Damrauer, L., Milburn, R. M.: J. Amer. Chem. Soc. *90*, 3884 (1968); Damrauer, L.: Dissertation Abstr. *30B*, 2066 (1969).
72) Bunton, C. A., Carter, J. H., Llewellyn, D. R., O'Connor, C., Odell, A. L., Yih, S. Y.: J. Chem. Soc., *1964*, 4615.
73) Odell, A. L., Olliff, R. W., Seaton, F. B.: J. Chem. Soc., *1965*, 2280.
74) Barton, D., Harris, G. M.: Inorg. Chem. *1*, 251 (1962).
75) Muetterties, E. L., Guggenberger, L. J.: J. Amer. Chem. Soc. *96*, 1748 (1974).
76) Larsen, E., La Mar, G. N., Wagner, B. E., Parks, J. E., Holm, R. H.: Inorg. Chem. *11*, 2652 (1972).
77) Kepert, D. L.: Inorg. Chem. *11*, 1561 (1972).
78) Stiefel, E. I., Brown, G. F.: Inorg. Chem. *11*, 434 (1972).
79) Healy, P. C., White, A. H.: J. Chem. Soc. Dalton, *1972*, 1163.
80) Gillum, W. O., Wentworth, R. A. D., Childers, R. F.: Inorg. Chem. *9*, 1825 (1970).
81) Wentworth, R. A. D.: Coord. Chem. Rev. *9*, 171 (1972/73).
82) Pignolet, L. H.: Inorg. Chem. *13*, 2051 (1974).
83) Day, V. W., Hoard, J. L.: J. Amer. Chem. Soc. *92*, 3626 (1970).
84) Fackler, J. P., Avdeef, A., Costamagna, J.: Proc. XIVth Int. Conf. Coord. Chem., *1972*, 589.
85) Taube, H.: Chem. Revs. *50*, 69 (1952).

Received June 14, 1974

A Theoretical Approach to Heterogeneous Reactions in Non-Isothermal Low Pressure Plasma

Dr. Stanislav Vepřek

Institute of Inorganic Chemistry, University of Zürich, Switzerland.

Contents

I. Introduction

The Chemical effects produced by various kinds of electrical discharge have attracted the attention of chemists since the middle of the last century. A large number of papers are available on homogeneous reactions under plasma conditions[1-5], but only little work was done on gas-solid reactions prior to 1960. Studies on the reduction of certain oxides, sulphides and halides[6,7] and on the formation of hydrides of tin, arsenic, antimony, etc.[8] are noteworthy. Anomalous sputtering of certain elements, such as As, Sb, Bi, Se and Te in a hydrogen glow discharge appeared to be due to chemical transport taking place by formation and subsequent decomposition of volatile, unstable hydrides[9].

The rapid development of solid state physics and technology during the last fifteen years has resulted in intensive studies of the application of plasma to thin film preparation and crystal growth The subjects included the use of the well known sputtering technique, chemical vapour deposition ("CVD") of the solid in the plasma, as well as the direct oxidation and nitridation of solid surfaces by the plasma. The latter process, called "plasma anodization"[10], has found application in the preparation of thin oxide films of metals and semiconductors. One interesting use of this technique is the fabrication of complementary MOS devices[11]. Thin films of oxides, nitrides and organic polymers can also be prepared by plasma CVD.

These applications will not be discussed here; the reader is referred to review articles and books[10, 12-14] and to more recent papers[15-19].

All the work just mentioned is rather empirical and there is no general theory of chemical reactions under plasma conditions. The reason for this is, quite obviously, that the ordinary theoretical tools of the chemist, — chemical thermodynamics and Arrhenius-type kinetics — are only applicable to systems near thermodynamic and thermal equilibrium respectively. However, the plasma is far away from thermodynamic equilibrium, and the energy distribution is quite different from the Boltzmann distribution. As a consequence, the chemical reactions can be theoretically considered only as a multichannel transport process between various energy levels of educts and products with a nonequilibrium population[20, 21]. Such a treatment is extremely complicated and — because of the lack of data on the rate constants of elementary processes — is only very rarely feasible at all. Recent calculations of discharge parameters of molecular gas lasers may be recalled as an illustration of the theoretical and the experimental labor required in such a treatment[22, 23].

A similar theoretical treatment of heterogeneous reactions under plasma conditions is even more complicated, owing to our poor knowledge of chemical interactions between plasmas and solids. Moreover, such an approach is hardly suited to the needs of a chemist in the laboratory, who is interested in preparative solid state chemistry, and who would prefer a reasonably simplified theoretical approach, which would give him a rough idea of the steady state chemical composition of the plasma and its dependence on the basic parameters of the discharge. Such an approach is reviewed in this article.

There are many different kinds of discharge plasmas[4,24-27] and it is not possible to develop a theoretical model applicable to all of them. We must therefore restrict ourselves to some particular kind of plasma, the choice of which will be determined by its practical applications in preparative solid state chemistry. This aspect will be discussed in Part II, of this article.

II. Discharges Suitable for Chemical Vapour Deposition

The physical plasma is either a partially or a fully ionized gas, which is macroscopically neutral. This means that the concentration of positively charged species is equal to that of the negatively charged ones. There are a number of different kinds of plasmas: In nature, there are the thermal plasmas of the stars, the interstellar plasmas, the lightning, the corona, etc. Various types of plasma can also be produced in the laboratory[24−27]. We shall now consider some of them from the aspect of their applicability to CVD[28].

A. The Arc Discharge

In an arc discharge at atmospheric pressure, the neutral gas temperature reaches a value of several thousand °K. No solid can exist at such high temperatures and, therefore, a deposition of solid can take place only in the outer region of such discharges. Fast quenching processes have to be accounted for in any theoretical treatment[20,21]. This plasma was used as a source of heat in a modification of the Verneuil method of crystal growth[13], powder spraying, etc.[14].

B. The Corona, the Low Current Glow Discharge and the Afterglow

The corona, the low current glow discharge and the afterglow are among the weakest discharges which can be produced. The neutral gas temperature can be adjusted within wide limits, either by heating or cooling of the discharge tube. The internal energy of these plasmas is close to the equilibrium value, but there are highly reactive species like atoms, ions and free radicals, the presence of which can result in high overall reaction rates, even at low concentrations. Such plasmas can be used as catalysts in systems in which a particular chemical reaction, though thermodynamically possible, does not take place because of the high activation energy without the plasma. Except in the electrode regions, the chemical equilibrium is determined by the value of the standard Gibbs energy of the reaction, $\Delta G°$, while the reaction rate is controlled by the concentration of the active species. The well known Boudouard's reaction $C(s) + CO_2(g) \rightleftharpoons 2\,CO(g)$ was studied as an example of such a reaction[28].

The theoretical treatment becomes more complicated if the electrode regions are taken into account − for example, when the solid is deposited on a thin wire acting as the central electrode of a corona discharge. Such an arrangement was used by Wales[29] in the deposition of boron by decomposition of boron tribromide. The decomposition was shown to take place in the proximity of the central wire electrode. The corona discharge is most intense near the wire and the decomposition is actually physical in nature. It takes place by electron impact or by dissociatve deexcitation of the excited species.

C. The High Current, Low Pressure Discharge

Finally we shall consider the plasmas of intense low pressure discharges. These plasmas can be produced either with direct or alternating current under pressures between 0.1 and several millimeters of mercury at electrical current between 0.1 A and several ampères[24–27]. The internal energy of these plasmas attains high values corresponding to an equilibrium temperature of several thousand degrees. However, the temperature of the neutral gas can be maintained at a much lower value, which is desirable for preparative purposes[28, 30–35]. These unique properties of intense low pressure discharges open new, attractive possibilities in preparative solid state chemistry.

However, such systems are very far away from thermodynamic equilibrium and their theoretical treatment is, therefore, extremely complicated. The distribution of energy cannot be described either by Boltzmann's distribution function or as small deviations from it (linear region). For example, the mean electron energy corresponds to a temperature of several tens of thousands °K, whereas the neutral gas temperature is lower by two or three orders of magnitude. Most of the earlier theoretical treatments are based on the simplifying assumption that the electron energy distribution is Maxwellian. It has been recently shown, however, that this is true only at high discharge currents, when the relatively high degree of ionization (10^{-5}–10^{-4}) and the low electric field strength favor an effective mutual electron-electron Coulomb interaction[46]. It should also be pointed out that molecular plasmas, which are of primary interest for CVD, are much less well understood than the plasmas of rare gases.

In such plasmas, the electron concentration does not exceed 10^{12} cm^{-3} and the electron temperature is typically around 2×10^4 °K[37]. Special attention must be paid to dissociation processes. At a discharge current of about 1 A the degree of dissociation reaches a value of several tens of percent in molecular gases which corresponds to an energy of several tens of kilocalories per mole. This is the largest part of the internal energy of such a plasma. The reason for this high degree of dissociation is the rapid dissociation of molecules by electron impact and by energy transfer from the excited species, as compared to the slow recombination of atoms on the walls of the discharge tube.

As an example, Table 1 gives a comparison of energy values which are stored in various subsystems of the plasma of a nitrogen high frequency discharge at 1 mm Hg and a medium value of the high frequency current[38]. It is seen that the largest part of the internal energy of the plasma is stored in the dissociation of molecules into atoms, because of the high degree of dissociation α.

Due to the prevalence of the dissociative processes, the chemical composition of the plasma is quite different from that of the corresponding heterogeneous systems

Table 1. Energy content of various subsystems of the low pressure plasma. Example: Nitrogen high frequency discharge, 27 MHz, $p. D.$ = 1 Torr cm, I/D = 1 A cm^{-1}, degree of dissociation $\alpha \approx 0.6$, T_t = 873°K

	kcal mole^{-1}
$E_{translational}$	= 1.8
$E_{rotational}$	$\doteq$ 1.8
$E_{vibrational}$	$\geqq$ 1.8
$E_{dissociation}$	$\doteq$ 135
$E_{metastable}$	$\leqq$ 6
$E_{short\ lived\ states}$	$\doteq$ 0.3
E_{ions}	$\doteq 10^{-3}$
$E_{electrons}$	$\doteq 10^{-3}$

under conditions of thermodynamic equilibrium. Thus, for instance, besides oxygen, only a few percent of carbon monoxide have been found in the plasma of the heterogeneous system solid carbon/oxygen between 300 °C and 700 °C[39]. On the other hand, the gas phase in thermodynamic equilibrium contains only CO_2 below 400 °C, a mixture of CO_2 and CO at intermediate temperatures and only CO (no free oxygen) above 1000 °C (see, *e.g.*,[40]). Similarly, no methane was found in the plasma of the C/H_2-system at low temperatures[39] even though it is the dominant component of the gas phase in thermodynamic equilibrium[40].

It is evident from these two examples, that the chemistry of low pressure plasmas is fundamentally different from the chemistry of ordinary systems. Unfortunately, the actual chemical composition of these plasmas is not known in most cases. The high degree of dissociation indicates that only simple (diatomic?) species which are generated by fast reactions on the walls of the discharge tube may survive in the plasma. This fact has to be kept in mind in any theoretical consideration of chemical processes taking place in such systems.

III. Chemical Transport in Low Pressure Plasma

A. General Considerations

Chemical transport of solids is a well known preparative technique. As was pointed out by Schäfer[41], information on thermodynamic properties of heterogeneous systems can also be obtained from experiments involving chemical transport. In particular, the dependence of chemical equilibrium of a heterogeneous reaction of the type

$$A(s) + B(g) \rightleftharpoons C(g) + D(g) \tag{1}$$

on temperature, pressure, etc. can be investigated in this way.

S. Veprek

The use of low pressure plasma for chemical transport was proposed by Hauptman in 1965[42]. Since then, this idea has been developed in a series of papers[28,30–35] It was shown that relatively simple chemical transport experiments under plasma conditions can reveal the fundamental nature of the dependence of the steady state composition of the plasma on the discharge parameters. Transport experiments in open (gas flow) systems yield information on the amount of solid A(s) which is chemically diluted in the gas phase due to the formation of the volatile compounds [*e.g.* C(g) – see Eq. (1)].

We shall consider the transport of carbon in hydrogen, oxygen and nitrogen as an illustration of the experiment and of the theory. Similarly to the temperature gradient which is used in ordinary transport systems[41], a gradient of plasma energy has to be produced. As pointed out above, the internal energy of the plasma is not determined by the temperature of the neutral gas. The energy gradient can be established, for example, by changing the discharge current density; this may be achieved by a change in the diameter of the discharge tube.

B. The Transport of Carbon

The experimental arrangement for the transport of carbon in hydrogen and oxygen is schematically shown in Fig. 1 (for further details see[28, 35]). The carbon charge is placed in the low energy zone (E_1) and a high frequency discharge is started at a total pressure of hydrogen between 0.4–0.8 mm Hg and a gas flow rate of about 100 mm Hg cm^3/sec. If the density of the high frequency current in the high energy zone (E_2) has reached a critical value, carbon is deposited in this zone.

In a similar manner carbon can also be transported in an oxygen plasma. However, the density of the electric current in the high energy zone (E_2) must reach a much higher value than for transport in hydrogen.

On the other hand, carbon can be transported in nitrogen in the direction of decreasing plasma energy, $E_2 \rightarrow E_1$[35]. The carbon charge is placed in the high energy zone and the gas is made to flow in the direction from left to right (see Fig. 1).

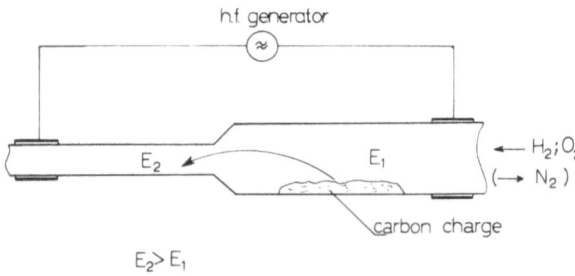

Fig. 1. Discharge tube with two zones of different plasma energies $E_1 < E_2$. Solid carbon is transported into the high energy region in hydrogen and oxygen $(E_1 \rightarrow E_2)$ and in the opposite direction in nitrogen $(E_2 \rightarrow E_1)$.

144

These experiments show that with increasing plasma energy the chemical equilibrium of the reaction (1) is shifted to the right in the C/N_2-system (transport direction $E_2 \rightarrow E_1$) whereas it is shifted to the left in the C/H_2- and C/O_2-systems (transport direction $E_1 \rightarrow E_2$). The mechanism of the transport was previously discussed for the C/N_2-system[28, 35, 43] and C/H_2- and C/O_2-systems[28, 39, 44]. It was shown that the transport involves formation and subsequent decomposition of CN, simple hydrocarbon radicals and CO in the systems C/N_2, C/H_2 and C/O_2 respectively.

Moreover, the systems obey the Le Châtelier-Braun's principle which indicates, that they are stable as regards their chemical composition. We shall discuss this point in the next section.

IV. Theoretical Model

In this section we shall recall some of the basic concepts and discuss the problem of stability of non-equilibrium systems. This will lead us to an idealized, simple theoretical model of low pressure plasma which can be mathematically treated.

A. Non-Equilibrium Systems

In the macroscopic theory, a state of a physical (chemical) system is described by a set of thermodynamic parameters ξ_j, $j = 1, \ldots, n$. These parameters and their derivatives determine the values of the thermodynamic fluxes J_i and forces X_i, $i = 1, \ldots, r$[45, 46].

All forces X_i and fluxes J_i vanish if thermodynamic equilibrium has been attained. The equilibrium values of the thermodynamic parameters are denoted as ξ_j^o.

By using external reservoirs, some of these parameters can be kept at values different from those of thermodynamic equilibrium, $\xi_j \neq \xi_j^o$, $j = 1, \cdots, m \leqslant n$. As a result, a non-equilibrium state arises, which is characterized by nonvanishing values of some fluxes J_i, $i = 1, \cdots, s \leqslant r$ and of the corresponding forces X_i. Examples of such processes are: diffusion and related effects, Peltier effect, etc.[45, 46]. Such a state can either be stationary or time-dependent, stable or unstable.

Most of the non-equilibrium systems were studied in linear approximation, i.e. near thermodynamic equilibrium; $\xi_j = \xi_j^o + \delta \xi_j$, $\delta \xi_j \ll \xi_j^o$. All such systems are stable[45] and are susceptible to at least an approximate mathematical treatment[46]. Such a treatment becomes highly complicated for systems which are out of the linear region, like biological systems and various physical and chemical systems with large gradients of the parameters ξ_j. In such systems instabilities may occur and new dissipative structures can be formed[45, 47]. Let us emphasize that such systems are "open", and they can exist only due to an exchange of energy, matter, etc. with some external reservoirs.

B. Stability Criterion

The condition of stability of these non-equilibrium states is given by the Glansdorff-Prigogine criterion[45]. The equivalence of this criterion with the extended form of Le Châtelier-Braun principle for systems which are far away from thermodynamic equilibrium[45], yields a convenient experimental criterion of stability: An open system which is far away from thermodynamic equilibrium is stable, if a change of some of the parameters ξ_j governing the particular non-equilibrium state results in a change of this state according to the principle of Le Châtelier-Braun.

As was pointed out in the last section, experiments on chemical transport of solids in intense low pressure discharges showed that the chemical composition of the plasma obeys this principle as regards the change of the internal energy. Blaustein and Fu[48] illustrated the validity of Le Châtelier Braun principle as regards the changes in the initial chemical composition. We can also conclude that these systems are stable as far as their chemical composition is concerned (for further discussion see[38, 43, 44]).

This obiously does not exclude the possibility of oscillation phenomena such as ionization waves, etc., which are among the most frequently observed instabilities in d.c. glow discharges[49]. Any allowance for these phenomena in the theoretical treatment of chemical processes would introduce major complications and, we shall therefore ignore them in the first approximation. Such a simplification is justified, because the relaxation processes related to the ionization waves are usually much faster than chemical processes. Therefore, the oscillations take place on a time scale which is different from that of the chemistry. Moreover, the waves arise in a d.c. discharge rather than in a high frequency discharge, which was used in most experiments on chemical transport.

In order to develop a simple theoretical model of the plasma under consideration, we must now give a microphysical interpretation of the stability criterion. Unfortunately, very little work has been done on the subject. A detailed stochastic treatment of several simple models has shown that stability will be ensured if the time scales associated with the fluctuations in the system are much shorter than the time scales associated with the outside world[45, 50) a)]. This condition is similar to that for local thermodynamic equilibrium[45, 52].

C. Statistical Model of Low Pressure Plasma

It is plausible to assume that the stability of the chemical composition of the plasma with respect to a disturbance from the "outer world" originates in a similar separation of the time scale of the chemical processes in the plasma from that of the interaction of the plasma with the electron gas and with the environment.

---

a) As pointed out in Ref.[51], this condition does not preclude instabilities on a macroscopic scale. Plasma instabilities such as standing striations might be a counterpart of the chemical instabilities discussed in[51].

The chemical reactions which have to be considered involve atoms, simple (diatomic?) molecules and radicals formed in high concentrations (10^{15}–10^{16} cm^{-3}). Numerical estimation shows that these reactions predominate over ion-molecule reactions in the type of plasma under consideration (see e.g.[28, 44]). Therefore, the latter reactions can be omitted here.

As the next step, we have to compare the relaxation times of various kinds of energy transformation. The following processes have to be considered (Table 2): energy exchange within the system of long-lived neutral species (atoms, radicals, . . .) due to fast chemical reactions ($\tau_{chemistry}$), energy flow from the electron gas into this system ($\tau_{excitation}$) and the energy exchange between the system and the environment (τ_{energy}). These processes were discussed in[38, 43, 44] and the typical values of the relaxation times are summarized in Table 2. In addition, there are also the relaxation times of energy equilibration in the translational rotational and vibrational degrees of freedom. The last-named equilibration is related to the relaxation of high excited vibrational states in chemically reacting systems. The lower states ($v = 1, 2$) of homopolar molecules are deactivated rather slowly, but they do not significantly contribute to the total energy of the plasma under consideration (see also[38, 43])[b].

Table 2. Relaxation times for various processes in the low pressure plasma

	sec
$\tau_{chemistry}$	10^{-4}–10^{-5}
$\tau_{excitation}$	10^{-1}–10^{-3}
τ_{energy}	10^{-1}–10^{-2}
$\tau_{translational}$	10^{-7}
$\tau_{rotational}$	$5 \cdot 10^{-7}$–10^{-7}
$\tau_{vibrational}$	10^{-5}–10^{-6}

It is seen from Table 2 that

$$\tau_t \leqslant \tau_r \leqslant \tau_v \ll \tau_{chem.} \ll \tau_{exc.} \lesssim \tau_{en.} \tag{2}$$

The inequality (2) considerably simplifies further treatment and is the starting point for a simple statistical model of the plasma[43].

The approach of a system characterized by inequality (2) to a steady state was previously discussed[38], and the properties of the steady state have been studied by methods of statistical physics[43]. We shall briefly summarize the results.

If the wave functions of individual species are separable (Born-Oppenheimer approximation), and if there is a weak energy coupling between the system of long-lived, chemically reacting particles with other degrees of freedom, the total probability distribution function $p(t, r, v, e)$ of the system can be separated:

---

[b] At higher pressures, however, the situation may well be different[53, 54].

$$p(t, r, v, e) = p(t) \cdot p(r) \cdot p(v) \cdot p(e). \tag{3}$$

Due to the above-mentioned weak coupling and to the inequality (2), the distribution of translational, rotational and, to an approximation also of vibrational degrees of freedom — $p(t), p(r)$ and $p(v)$ respectively — can be satisfactorily approximated by the Boltzmann functions $\exp(-\beta_\kappa \epsilon_\kappa)$ with the corresponding temperatures

$$T_t = (k\beta_t)^{-1} \lesssim T_r = (k\beta_r)^{-1} \lesssim T_v = (k\beta_v)^{-1}. \tag{4}$$

Here, β_κ are the modules of canonical distributions and ϵ_κ are the energy values.

Special attention has to be paid to the electronic degrees of freedom. A set of energy levels of reaction educts and products which are strongly coupled due to fast chemical reactions and energy transfer was considered in the simple model[43]. Because of the weak coupling of this system to the "outer reservoirs",

$$\tau_{\text{chem.}} \ll \tau_{\text{exc.}} \lesssim \tau_{\text{en.}} \tag{2a}$$

the Boltzmann energy distribution within these electronic states results. The corresponding temperature is:

$$T_e = (k\beta_e)^{-1} \gg T_{t, r, v} \tag{5}$$

[see (4)]. T_e is not identical with the temperature of the electron gas [see (2a)]. If $\tau_{\text{exc.}} < \tau_{\text{en.}}$, the energy coupling of the system to the environment is the slowest process, which determines the overall relaxation time of the energy of the system.

Such a steady state is evidently stable. Moreover, due to the inequality (2a), the system of long-lived, chemically reacting particles "forgets" its mode of reception of the energy from the electrons, and the mode of energy dissipation into the environment. The energy within the system will be distributed according to the particular chemical reactions involved. This means that all chemical fluxes vanish in the steady state, although there still remain nonvanishing fluxes of energy. We have a chemical equilibrium state which is far removed from the thermodynamic equilibrium (for further discussion see[44]).

The chemical equilibrium constant of an ordinary chemical reaction

$$A + bB \rightleftharpoons C \tag{6}$$

is given in terms of partition functions in the same form as in the thermodynamic equilibrium[43, 55]

$$K = \frac{q_C}{q_A \cdot (q_B)^b} \tag{7}$$

However, the partition functions have another form which, for our particular model, is given by[43]:

$$q_x = \sum_e g_e^x \, f^{\prime x}(\epsilon_e^x) \, \exp(-\beta_e \epsilon_e^x) \cdot \underset{(t,\,r,\,v)_e}{\Sigma} \underset{(\kappa\,=\,t,\,r,\,v)}{\Pi} g_{\kappa e}^x \, \exp(-\beta_\kappa \epsilon_{\kappa e}^x). \qquad (8)$$

Here, the index e stands for electronic, and the index κ for either translational, rotational or vibrational degrees of freedom, the quantities β are the modules of canonical distributions [see (4) and (5)], and ϵ are the energy levels. The function $f^{\prime x}(\epsilon_e^x)$ is an expression of the fact that only certain electronic levels of individual chemical components — the long-lived, chemically interacting ones — in the system are considered.

In case of a heterogeneous reaction

$$A(s) + bB(g) \rightleftharpoons C(g), \qquad (6a)$$

the following species should be taken into account: molecules and atoms of the transporting agent B(g) in their ground and electronically excited metastable states, and the volatile products C(g) of fast reactions of gaseous species B(g) with the solid A(s). If the solids considered are not vaporized at the temperature T_t in the plasma ($T_t \lesssim 1300\,°K$), only the electronic ground state of A(s) appears in the respective partition function [see Eq. (29) in Ref.[43]].

Figure 2 illustrates the effect of the change in the internal energy of the plasma on the chemical equilibrium composition. The high energy of the plasma can be stored only in the electronic energy levels of the gaseous species B(g) and C(g) given in Fig. 2. The mean energy stored in the translational, rotational and vibrational degrees of freedom is much smaller than that of the long lived metastables, atoms, radicals, etc. (see Table 1). Consequently, the former degrees of freedom do not significantly contribute to the energy diagramm (Fig. 2)[c].

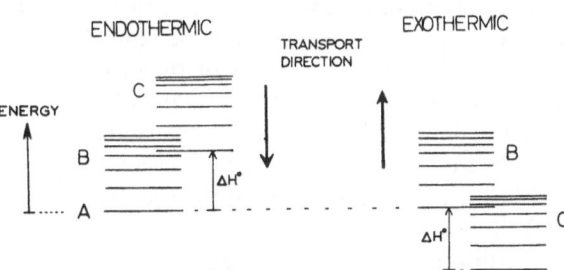

Fig. 2. Electronic energy levels in a chemically reacting system. $\Delta H°$ is the reaction enthalpy

Consider first a strong endothermic reaction, $\Delta H° \gg RT_t$ (see Fig. 2). Starting from a weak plasma, the energy is stored in the electronic levels of the transport agent B(g), and only if the high energy levels of B(g) are populated ($\epsilon_e^B \gg \Delta H°$), are the gaseous reaction products significantly available in the plasma. The solid A(s) is transported in the direction of decreasing plasma energy in such systems (e.g. the transport of carbon in nitrogen).

---

[c] For more details see also Eq. (33) in [43] and the discussion of this equation.

An opposite direction of the transport is observed in systems which are strongly exothermic without the plasma, $-\Delta H° \gg RT_t$. However, the situation in this case is more complicated.

In a number of such systems the reaction products are polyatomic molecules containing three or more atoms. With increasing plasma energy these molecules become dissociated to an increasing degree and species containing smaller numbers of atoms appear in the gas phase. The formation of these species is endothermic as regards the thermodynamically stable products of the reaction (6a) without the plasma. However, their formation can be exothermic if the atoms of the transporting agent B(g) react with the solid A(s). Consequently, we may consider only the simplest, diatomic species in the plasma[56].

$$A(s) + B(g, \text{atomic}) \underset{r_2}{\overset{r_1}{\rightleftharpoons}} AB(g) \qquad \Delta H° < 0 \qquad (6b)$$

If $-\Delta H° \gg RT_t$ for reaction (6b), the deposition of the solid is strongly endothermic. The activation energy is frequently found to be very low for the forward reaction, $E_1 (\text{act.}) \approx RT_t$, but it is very high for the reverse reaction:

$$E_2(\text{act.}) = E_1(\text{act.}) -\Delta H° \gg RT_t.$$

The formation of the species AB(g) can take place by thermal collision of an atom of B(g) with the solid surface, but only an energy transfer from a gaseous species carrying an energy of several electron volts can result in the decomposition of AB(g) at a rate r_2 which would be comparable with r_1.

This question was discussed in[28] and, in more detail in[44]. It was shown that the most efficient process of decomposition of the species AB is the dissociative de-excitation of metastable atoms of B, if their energy is higher than the dissociation energy of AB (see for example the C/H_2-system in Section V).

If, on the other hand, the excitation energy of the metastable is lower than the dissociation energy of the species AB, the latter can only be decomposed either by dissociative de-excitation of short-lived electronically excited states, or by ion-molecule reactions, or by electron impact. Much higher plasma energy values are necessary to produce a high rate of decomposition of the species AB in such systems[28, 44] (e.g. C/O_2, TiN/Cl_2).

After the species AB(g) has been decomposed by one of these processes, the solid A(s) will be deposited for the following reasons: the energy necessary for the sublimation of the solid has to be made available to the vibrational degrees of freedom of the solid. These are nearly in thermal equilibrium with the translational degrees of freedom of the gas and their mean energy is several kcal/mole at $T_t \lesssim$ 1300 °K. Therefore, the value of the partial pressure p_A of component A corresponds to that of thermodynamic equilibrium at T_t, and is negligible when the enthalpy of sublimation is much higher than the mean thermal energy. Obviously, such solids are the only ones here considered, because there is rarely need to transport the solid by the use of plasma if it evaporates at a low temperature.

So far we have considered an idealized model possessing most of the significant properties of the plasma of intense low pressure discharges. The choice of a simple model has made it possible to use an exact mathematical treatment which gives a transparent relationship between the physical assumptions and the behavior of such systems. Real systems display the properties of the theoretical model as exactly as they conform to the assumptions which underlie the inequality (2). The validity of (2) must be checked in any application of the theory to some particular system. In the next section, we shall put forward several fundamental ideas on this point by discussing the individual terms of the inequality (2) one by one.

V. Real Systems

The relaxation times τ_t and τ_r are evidently the shortest in the plasma under consideration and $\tau_t \lesssim \tau_r \ll \tau_{chem}$ can be assumed to be valid in all real systems.

On the other hand, the relaxation times of lower vibrationally excited states may be comparable with that of chemical processes. The short relaxation times τ_v indicated in Table 2 refer to the relaxation of higher vibrational levels in chemically reacting systems. The lower states ($v = 1, 2, 3$) can be more densely populated, and $T_v > T_t$. If the energy of formation and/or dissociation of the long-lived chemically reacting species is much larger than the vibrational energy quantum, the low vibrational states will not significantly affect the chemical equilibrium composition. Generally, a higher vibrational temperature will result in a lower concentration of the species containing more than one atom.

The most serious limitation of the applicability of the theoretical model to real systems is the validity of inequality (2a). The condition

$$\tau_{chem.} \ll \tau_{en} \qquad (2a')$$

means that the heterogeneous recombination of atoms and/or the de-excitation of metastables on the surface of the solid A(s) [see Eq. (6a)] is much slower than the chemical reaction. We must consider endothermic and exothermic systems separately.

A. Endothermic Systems

Such systems may be illustrated by the system[35] C/N_2. There are two competitive processes for the atoms on the surface: recombination (9a) and reaction with the solid (9b).

$$N \ + N \ \to N_2(g) \qquad (9a)$$
$$C(s) + N \ \to CN(g) \qquad (9b)$$

The former, (9a), represents energy dissipation into the environment ($\tau_{en.}$), the latter (9b) corresponds to the energy exchange within the system ($\tau_{chem.}$). The recombination is usually faster at low temperatures, while the reaction with the solid becomes significant at $800-1000\,^\circ C$ (see discussion in Ref.[28, 35, 43]). This is also apparent from the measured surface temperatures of different solids in a high frequency nitrogen discharge[59]. Very little is known about the de-excitation of the metastables on the surface. There seems to be some similarity with the recombination of atoms[60-62], but a more thorough study of these processes is desirable. Nevertheless, we can restrict our considerations to the atoms, because their concentration is several orders of magnitude higher than that of atomic metastables in the plasma.

The translational temperature T_t plays an important kinetic role. At high temperatures chemical reactions are fast, and − in view of the decreasing rate of surface recombination of atoms − the energy exchange of the system with the environment becomes slower. Consequently, the theoretical model can be applied to such systems (for comparison see Table 1 in[43]). The actual equilibrium concentration of the volatile reaction products − CN in the present case − may be reduced by dissociative de-excitation of electronically excited species (cf. also the system C/H_2).

B. Exothermic Systems

As an example of exothermic systems we shall consider the system C/H_2[28]. The corresponding energy level diagram, which is a counterpart of Fig. 2, is given in Fig. 3. It has been shown that there are only simple hydrocarbon radicals in the plasma under conditions of chemical transport[39]. The deposition of carbon, *e.g.* CH(g) →

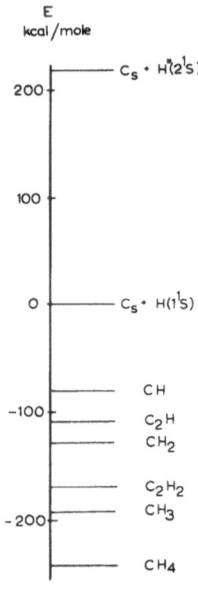

Fig. 3. Energy level diagram for the system carbon/hydrogen

C(s) + H, takes place at a temperature of about 950 °C. As in the previous case, the reaction of hydrogen atoms with solid carbon is fast at this temperature[67]. The decomposition of the hydrocarbons like CH-radical takes place by dissociative de-excitation of the hydrogen metastables[44]:

$$CH(g) + H\,(2^1S) \;\rightarrow C + H + H \tag{10}$$

Carbon atoms are deposited on the glass wall of the discharge tube because of the very low pressure of carbon vapour at the neutral gas temperature $t_t \lesssim 950\,°C$. Thus, carbon is transported into the high energy zone.

However, the deposition of carbon does not take place on metal surfaces, such as platinum. The reason for this effect is the high catalytic efficiency of pure metals for surface recombination of atoms and metastables, which leads to a decrease of the plasma energy near the surface. Consequently, inequality (2a') is not fulfilled there (cf.[28]).

An electron concentration $n_e \leq 5 \times 10^{11}\,cm^{-3}$ and electron temperature $T_e \approx$ 2 eV was measured in high frequency discharges in molecular gases such as H_2, CH_4, NH_3, etc.[37a]. The power of about $10\,W\,cm^{-3}$, dissipated in these discharges, was rather higher than in our own systems and, therefore, the above values of n_e and T_e are the highest ones which can be expected in systems such as C/N_2 and C/H_2. Thus, the inequality

$$\tau_{chem.} \ll \tau_{exc.} \tag{2a''}$$

is well satisfied in these systems (see also[43, 44]).

A different situation arises when the excitation energy of the metastables is lower than the dissociation energy of the corresponding species, as is the case for example, in the system C/O_2 [28]. Carbon is transported according to the reaction (see also Ref.[39]).

$$C(s) + O \rightleftharpoons CO \tag{11}$$

in the direction of increasing plasma energy. A much higher density of h.f. current is required in the deposition zone than in the system C/H_2, because carbon monoxide can be decomposed only by electron impact, ion-molecule reactions and/or dissociative de-excitation of higher, short-lived excited states. Consequently, the interaction of the system of long-lived chemically reacting species with the electron gas has to be taken into account.

Similar conditions obtain in the systems TiN/Cl_2 and AlN/Cl_2 [33]. Figure 4 shows the respective energy level diagrams[d]. Without the plasma, the chemical equilibrium is shifted far towards the stable chlorides. In the plasma, an increase of the internal energy leads to an increase of dissociation with the formation of simpler species till free metal atoms are formed in the gas phase. As in the system C/H_2, these atoms must condense on the wall of the discharge tube. At the same time they may react with nitrogen atoms with formation of the solid nitride. Because the dissociation energy of TiCl and AlCl is higher than the excitation energy of the metastable $N(^2D)$ and $N(^2P)$, the deposition of nitrides may be expected to take place only at a relatively high density of the high frequency current. This conclusion is confirmed by experimental results[33].

---

[d] For the corresponding energies of formation see[63, 64].

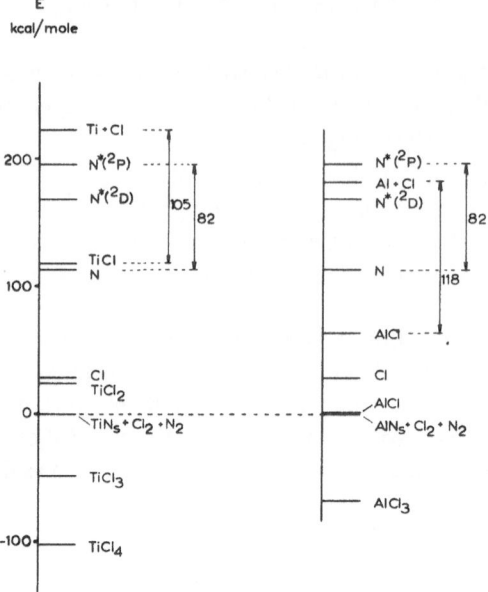

Fig. 4. Energy level diagram for the systems TiN/Cl_2 and AlN/Cl_2

Additionally, iodine may be employed instead of chlorine. The dissociation energies of the corresponding metal iodides[63] are lower than the excitation energy of the nitrogen metastables and the nitride should therefore be deposited at a much lower value of the plasma energy than that of the chlorine/nitrogen plasma. Therefore, the use of iodine would seem to be preferable for preparative purposes, but experimental results are still lacking.

C. Plasma Gradients

Finally, we shall briefly mention the problem of gradients of plasma parameters, including the degree of dissociation, electron concentration and temperature, etc. The assumption underlying the theoretical model is that it is possible to choose a small volume in which the parameters can be considered as constant, and which contains a number of particles which is large enough for statistical treatment.

In real systems we have to distinguish between two different types of plasma gradients: the radial gradients ("long range"-) , and the short-range-gradients near the surface of the solid.

The existence of the relatively small radial gradients in the plasma at a pressure of about 1 mm Hg involves no special problems concerning the applicability of the theory. It should, however, be kept in mind that in a wide tube the energy can be much higher in the middle than near the walls. In analogy to the transport taking place in the axial arrangement (Fig. 1), solids can be transported in radial direction if a discharge tube of a large diameter is used[33].

The short-distance gradients of plasma energy appear in the vicinity of solid surfaces possessing a high catalytic efficiency for the recombination of atoms and de-excitation of various excited species. These gradients may be very high and result in significant changes of the chemical processes. We have already mentioned that the deposition of carbon does not take place on such surfaces in C/H_2- and C/O_2-systems. Another effect was described in Ref.[38]. The titanium nitride crystals grown in a nitrogen/chlorine plasma[33] are simultaneously dissolved on the side adjacent to the substrate, if the energy in the deposition zone is relatively low. The reason for this effect is that the plasma energy decreases strongly in regions which are screened from the active discharge and, consequently, titanium nitride is not stable there.

Fig. 5. Molybdenum sheet recrystallized in nitrogen/chlorine plasma. Electron scanning microscope, 480x

Finally, a short distance transport resulting in recrystallization of the solid can take place owing to the existence of such gradients. Fig. 5 shows a molybdenum sheet which was recrystallized in a nitrogen/chlorine plasma[65]. The experimental arrangement was similar to that used in the synthesis of nitrides. The molybdenum sheet was inserted directly into the high energy zone (Zone II in Fig. 3, Ref.[33]). Fig. 5 shows the well-developed faces of small crystals. The recrystallization can be explained as follows (Fig. 6): in the beginning, there are small cracks in the sheet in which ordinary chemical reactions take place, i.e. the metal chlorides are formed. They diffuse out of the cracks into the high energy plasma where they are decomposed. Because of the relatively low temperature (about 1100 °C) the metal is deposited, and the most stable form — the crystals — is formed. It should be men-

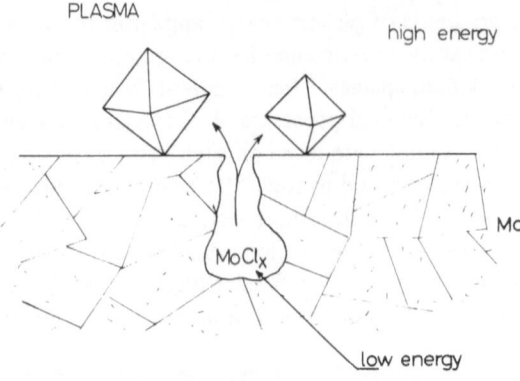

Fig. 6. Mechanism of recrystallization of molybdenum in nitrogen/chlorine plasma. A cross-sectional view of the sheet with the growing crystals

tioned that molybdenum nitride is not stable at this temperature[66] and the melting temperature of molybdenum is 2619 °C[63].

These results also support the theoretical explanation of the synthesis of titanium nitride and aluminum nitride given above. They show that metal chlorides can be decomposed and metals deposited in the intense low pressure plasma under similar conditions as those employed for the synthesis of nitrides. The interpretation of the experimental results given in Ref.[33] can now be considered as the initial step in the development of the ideas presented in this review (see also[30]).

VI. Conclusions

The purpose of the recent work just reviewed was to develop and verify a reasonably simplified theoretical approach to heterogeneous reactions in a nonisothermal low pressure plasma. With this purpose in mind, we first considered a simple statistical model of the plasma which has brought about a better understanding of the dependence of the chemical composition of the plasma on energy. Comparison of this model with several real systems which had been experimentally investigated illustrated the applicability of the theoretical ideas to such systems as well as their limitations.

The energy level diagrams, which can be compiled and used by any experimental chemist, represent a qualitative assessment of the dependence of chemical equilibrium composition of heterogeneous systems on the plasma parameters. The merit of such an approach is its simplicity.

On the other hand, the Boltzmann method of calculating the most probable distribution, used in the theoretical model, precludes an explicit consideration of actual values of transition probabilities (rate constants). This would only be possible if plasma reactions are considered as a multi-channel transport problem. However, the knowledge of a large number of various transition probabilities is necessary for such calculations. The determination of these constants is an important aspect of future work.

We also know very little about the chemical and physical interaction of the plasma with solid surfaces. Special attention must be paid to the problems of nucleation — heterogeneous as well as homogeneous — and to the mechanism of crystal growth under plasma conditions.

For the present, the theoretical approach described in this review makes it possible to predict the direction of chemical transport as well as the relative amount of plasma energy which is necessary for the transport to take place. Such considerations also allow a choice of the most suitable heterogeneous system for chemical evaporation and/or deposition of a particular solid. The exact plasma parameters have to be found experimentally. Nevertheless the theoretical approach is quite useful in planning experiments, as has been illustrated here for several systems.

The main subject of this review were the theoretical aspects of plasma CVD processes. The applications will be discussed in another paper. The results so far obtained show that the plasma of intense low pressure discharges offers new techniques of crystal growth. There are also other problems and tasks, which are just as important and which merit a more thorough study. These include for example, the application of plasma to thin film technology, the importance of which may be expected to increase in the future. The purification of materials by plasma transport can find practical uses, and many other applications will certainly be found.

The theoretical approach of this review is neither the only one possible nor the final one, and this contribution is more in the nature of a challenge to perform more sophisticated experimental work and to develop new ideals on the subject.

VII. References

1) Glocker, G., Lind, S. C.: The Electrochemistry of Gases and other Dielectric. New York: John Wiley 1939.
2) Steacie, E. W. R.: Atomic and Free Radical Reactions. New York: Reinhold 1946, 2nd Edition 1954.
3) McTaggart, F. K.: Plasma Chemistry in Electrical Discharges. Amsterdam: Elsevier 1967.
4) Venugopalan, M. (ed.): Reactions under Plasma Conditions, Vol. I and II. New York: John Wiley 1971.
5) Blaustein, B. D. (ed.): Chemical Reactions in Electrical Discharges. Adv. Chem. 80, Washington D. C.: Amer. Chem. Soc. 1969.
6) Bonhofer, K. F.: Z. physik. Chem. 116, 391 (1925); ibid, 113, 199 (1924).
7) Kroepelin, H., Vogel, E.: Z. anorg. allg. Chem. 229, 1 (1936).
8) Foresti, B., Mascaretti, M.: Gazz. Chim. Ital. 60, 745 (1930).
9) a) Günterschulze, A.: Z. Physik 36, 563 (1926).
   b) Günterschulze, A.: Vacuum 3, 360 (1953).
10) O'Hanlon, J. F.: J. Vac. Sci. Techn. 7, 330 (1970).
11) Micheletti F. B., Norris P. E., Zaininger K. H.: RCA Review 31, 330 (1970).
12) Mearns, A. M.: Thin Solid Films 3, 201 (1969).
13) Locker, L. D.: Materials Produced by Electrical Discharges. In: Modern Materials (ed. B. W. Gonser), Vol. 7, p. 89. London — New York : Academic Press 1970.
14) Gerdemann, D. A., Hecht, N. L.: Arc Plasma Technology in Material Science. Wien: Springer 1972.

15) Kassing, R., Deppe, H. R.: Thin Solid Films *13*, 27 (1972).
16) Tien, P. K., Smolinsky, G., Martin, R. J.: Appl. Optics *11*, 637 (1972).
17) Abe, H., Sonobe, Y., Enomoto, T.: Jap. J. Appl. Phys. *12*, 154 (1973).
18) Abe, H., *et al.:* Denki Kagaku *41*, 544 (1973).
19) Sanchez, D., Carchano, M., Bui, A.: J. Appl. Phys. *45*, 1233 (1974).
20) Polak, J.: Proc, Xth Int. Conf. Phenomena in Ionized Gases. Invited papers, p. 113, Oxford, England, 1971.
21) Polak, J., in: Reactions under Plasma Conditions (ed. M. Venugopalan), Vol. II, p. 141. New York: John Wiley 1971.
22) Novgorodov, M. Z., Sobolev, N. N.: Proc. 11th Int. Conf. Phenomena in Ionized Gases. Invited papers, p. 215, Prague, Czechoslovakia, 1973.
23) Nighan, W. L.: Proc. 11th Int. Conf. Phenomena in Ionized Gases. Invited papers, p. 267, Prague, Czechoslovakia, 1973.
24) von Engel, A.: Ionized Gases. Oxford, England: At the Clarendon Press 1955. 2nd Ed. 1965.
25) Francis, G.: Ionization Phenomena in Gases. London: Butterworth 1960.
26) Granowski, W. L.: Der elektrische Strom in Gas. Band I. Berlin: Akademie-Verlag 1955 (German transl.).
27) Granowski, W. L.: Der elektrische Strom in Gas. Vol. II, Moscow, Izdatelstvo Nauka 1971 (in Russian).
28) Veprek, S.: J. Crystal Growth *17*, 101 (1972).
29) Wales, R. D., in: Adv. Chem. (ed. B. D. Blaustein), Vol. 80, p. 198. Washington D. C.: Amer. Chem. Soc. 1969.
30) Veprek, S., Hauptman, Z.: Proc. IVth Yugoslav Symp. Phys. Ionized Gases. Herzeg Novi 1968.
31) Veprek, S., Hauptman, Z.: Z. anorg. allg. Chem. *359*, 313 (1968).
32) Veprek, S., Mareček, V.: Solid State El., *11*, 683 (1968).
33) Veprek, S., C., Brendel, Schäfer, H.: J. Crystal Growth *9*, 266 (1971).
34) Veprek, S., Oswald, H. R.: Meeting of the Swiss Cryst. Soc., Div. Cryst. Growth, Neuchâtel, Switzerland, October 1974. Z. allg. anorg. Chem. (in press).
35) Veprek, S.: Z. phys. Chem. N. F. *86*, 95 (1973).
36) Růžička, T., Rohlena, K.: Proc. 11th Int. Conf. Phenomena in Ionized Gases. Invited Papers, p. 61, Prague, Czechoslovakia 1973.
37) a) Rosskamp, G.: unpublished results, 1973.
    b) Janzen, G., *et al.:* Ber. Bunsen Ges. *78*, 440 (1974).
38) Veprek, S.: IEEE Trans. Plasma Sci. *PS-2*, 25 (1974).
39) Veprek, S., Cocke, D. L., Gingerich, K. A.: Chem. Phys. (in press).
40) Knippenberg, W. F., *et al.:* Philips Tech. Rundschau *28*, 143 (1967).
41) Schäfer, H.: Chemische Transportreaktionen. Weinheim; Verlag Chemie 1962, (Engl. transl.) London – New York: Acad. Press 1964.
42) Hauptman, Z.: unpublished, Prague 1965.
43) Veprek, S.: J. Chem. Phys. *57*, 952 (1972).
44) Veprek, S., Peier, W.: Chem. Phys. *2*, 478 (1973).
45) Glansdorff, P., Prigogine, I.: Thermodynamic Theory of Structure, Stability and Fluctuations. London: John Wiley 1971.
46) Groot, S. R., Mazur, P.: Non-equilibrium Thermodynamics, Amsterdam: North-Holland 1962.
47) Eigen, M.: Die Naturwissenschaften *58*, 466 (1971).
48) Blaustein, B. D., Fu, Y. C., in: Adv. Chem. (ed. B. D. Blaustein) Vol. 80, p. 259. Washington, D. C.: Amer. Chem. Soc. 1969.
49) Pekárek, L.: Proc. Xth Int. Conf. Phenomena in Ionized Gases. Invited papers, p. 365, Oxford, England 1971.
50) Nicolis, G., Babloyantz, A.: J. Chem. Phys. *51*, 2632 (1969).
51) Nicolis, G., Prigogine, I.: Proc. Nat. Acad. Sci. USA *68*, 2102 (1971).
52) Drawin, H. W., in: Reactions under Plasma Conditions (ed. M. Venugopalan), Vol. I, p. 53. New York: John Wiley 1971.
53) Cramarossa, F., Ferraro, G., Molinari, E.: J. Quant. Spectrosc. Radiat. Transfer *14*, 419 (1974).

54) Nelson, L. Y., Saunders, A. W. Jr., Harvey, A. B.: J. Chem. Phys. 55, 5126 (1971).
55) Manes, M., in: Adv. Chem. (ed. B. D. Blaustein) Vol. 80, p. 133. Washington D. C.: Amer. Chem. Soc. 1969.
56) For further discussion see [28, 35, 43]. For experimental examples see [39, 57, 58].
57) Neumann, W.: Jahresbericht der Akademie der Wissenschaften der DDR, Zentralinstitut f. Elektronenphysik, Berlin 1972.
58) Holzmann, R. T., Morris, W. F.: J. Chem. Phys. 29, 677 (1958).
59) Vepřek, S.: Ph. D.-Thesis. University of Zürich 1972.
60) Noxon, F. J.: J. Chem. Phys. 36, 926 (1962).
61) Haque, R., von Engel, A.: Int. J. Electronics, 36, 239 (1974).
62) Vidaud, P. H., von Engel, A.: Proc. Roy. Soc. A 313, 531 (1969).
63) JANAF Thermochemical Tables. US Natl. Bur. Std., 2nd ed., NSRDS-NBS 37, 1971.
64) Hasted, J. B.: Physics of Atomic Collisions. 2nd ed. London: Butterworths 1972.
65) Vepřek, S.: unpublished results.
66) Samsonov, G. V.: Nitridy, Kiev: Naukovaja Dumka 1969 (in Russian).
67) Recently, a detailed kinetic study of this system has been done by Coulon and Bonnetain: Coulon M., Bonnetain; L.: J. Chim. Phys. 71, 711 (1974), 71, 717 (1974), 71, 725 (1974).

Received November 21, 1974

C.K. Jørgensen:
Oxidation Numbers and Oxidation States

VII, 291 pages. 1969 (Molekülverbindungen und Koordinationsverbindungen in Einzeldarstellungen)
Cloth DM 68,−; US $27.90   ISBN 3-540-04658-5

For most chemists, formal oxidation numbers are a tool for writing correct reaction schemes more conveniently. However, recent progress in absorption spectroscopy of transition group complexes and the application of group theory and molecular orbital theory to chromophores, consisting of a central atom and a number of adjacent atoms belonging to the ligands, has made it possible to connect the classification of oxidation states with spectroscopic and magnetochemical results. Consequently, the apparent discrepancy between the integral number of electrons in the partly filled shell and the fractional atomic charges can be clarified in a satisfactory and rather unexpected fashion. However, under certain circumstances, the ligands are not innocent and do not allow the determination of oxidation states. The book treats many detailed problems, such as the phenomenological baricenter polynomial for electron configurations in monatomic entities, spin-pairing energy, the Madelung potential and the stabilization of definite oxidation states, the nephelauxetic effect, bonding in semiconductors, the question of back-bonding in carbonyl, olefin and hydride complexes,etc.

H.-H. Perkampus:
Wechselwirkung von π-Elektronensystemen mit Metallhalogeniden

Prices are subject to change without notice
Preisänderungen vorbehalten

64 Abbildungen. 37 Tabellen. XI, 215 Seiten. 1973 (Molekülverbindungen und Koordinationsverbindungen in Einzeldarstellungen)
Gebunden DM 78,−; US $32.00   ISBN 3-540-06318-8

Die Wechselwirkung von π-Elektronensystemen mit Metallhalogeniden (auch als Lewis- oder Ansolvosäuren zu bezeichnen) hängt von der Gegenwart von Protonen (aus protonenhaltigen Lösungsmitteln oder Verunreinigungen stammend) ab. Sind Protonen vorhanden, so resultieren sog. Proton-Additionskomplexe, fehlen Protonen, so entstehen π-Komplexe oder σ-Komplexe, letztere mit einer polaren Bindung zwischen Kohlenstoff und Metallatom. Diese drei Komplextypen werden in diesem Buch behandelt. Besonders ausführlich werden die Ergebnisse der Untersuchungen an den reinen binären Systemen π-Elektronensystem und Metallhalogenid dargelegt. Hier fehlte bisher eine umfassende Darstellung, die den Primärschritt der Wechselwirkung ohne Folgereaktion zum Gegenstand hatte. Zusammenfassend werden an den binären Systemen diskutiert: Dampfdruckmessungen, Röntgenstrukturuntersuchungen, UV-, IR-, Raman-, NMR-, NQR-, ESR- und DK-Messungen (Dipolmomente).

**Springer-Verlag
Berlin
Heidelberg
New York**

W. Schneider:
Einführung in die Koordinationschemie

38 Abbildungen. VIII, 173 Seiten. 1968
Gebunden DM 36,−; US $14.80   ISBN 3-540-04324-1

Organische Chemie in Einzeldarstellungen   15

A. Gossauer
DIE CHEMIE DER PYRROLE

17 Abbildungen
XIX, 433 Seiten. 1974
Geb. DM 158,—
US $64.80
ISBN 3-540-06603-9

Das Pyrrol und seine Derivate haben als technische Grundstoffe wie auch als Naturprodukte wachsendes Interesse gewonnen.

Diese Monographie ist eine umfassende Übersicht über die seit 1934 erschienene Literatur (ausgenommen Porphyrine). Bedingt durch die seitdem ständig wachsende Anzahl der Veröffentlichungen, die sich mit den physikalischen Eigenschaften dieser Verbindungsklasse befassen, weichen Konzeption und Gliederung dieses Buches von denjenigen des klassischen Werkes von H. Fischer und H. Orth grundsätzlich ab. Die Anwendung quantenmechanischer Rechenverfahren zur Deutung der Eigenschaften des Pyrrol-Moleküls wird im ersten Kapitel ausführlich erörtert. Die entscheidende Bedeutung der physikalischen Methoden zur Untersuchung der Konstitution und Reaktivität des Pyrrols und seiner Derivate ist durch zahlreiche tabellarisch geordnete Datenangaben, deren Interpretation im Text diskutiert wird, hervorgehoben. Dem präparativ arbeitenden Chemiker soll die Systematisierung der synthetischen Methoden bei der Suche nach der einschlägigen Literatur helfen: Ringsynthesen sind nach dem Aufbaumodus des Heterocyclus, die Einführung von Substituenten nach funktionellen Gruppen klassifiziert und anhand von Schemata übersichtlich zusammengefaßt worden. Bei der Zusammenstellung der Abbildungen wurden neben den trivialen Beispielen, die zum besseren Verständnis des Textes dienen, besonders jene Reaktionen ausgewählt, bei denen Pyrrole Ausgangsverbindungen zur Darstellung anderer Heterocyclen (Indole, Pyrrolizine, Azepine, u.a.) sind. Besondere Sorgfalt gilt der Beschreibung von Reaktionsmechanismen. (2621 Literaturzitate.)

Weitere Bände:

2   E. Clar: Aromatische Kohlenwasserstoffe. Polycyclische Systeme. Mit einem Geleitwort von J. W. Code. 2. verb. Aufl. 138 Abb. XXII, 481 S. 1952. DM 76,—; US $31.20 ISBN 3-540-01647-3

4   H. Henecka: Chemie der Beta-Dicarbonyl-Verbindungen. 10 Abb. VI, 409 S. 1950 Geb. DM 60,—; US $24.60 ISBN 3-540-01488-8

5   G. Schramm: Die Biochemie der Viren. 67 Abb. VIII, 276 S. 1954. DM 50,—; US $20.50 ISBN 3-540-01834-4

7   H. Meier: Die Photochemie der organischen Farbstoffe. 168 Abb. XVI, 471 S. 1963 DM 98,—; US $40.20 ISBN 3-540-03034-4

8   H. Suhr: Anwendungen der kernmagnetischen Resonanz in der organischen Chemie 123 Abb. VIII, 424 S. 1965. DM 94,—; US $38.60 ISBN 3-540-03380-7

9   E. Schmitz: Dreiringe mit zwei Heteroatomen. Oxaziridine, Diaziridine. Cyclische Diazoverbindungen. 5 Abb. XII, 179 S .1967 DM 78,—; US $32.00   ISBN 3-540-03946-5

10   J. Falbe: Synthesen mit Kohlenmonoxyd 20 Abb. VIII, 212 S. 1967. DM 64,—; US $26.30 ISBN 3-540-03947-1

11   K. D. Gundermann: Chemilumineszenz organischer Verbindungen. Ergebnisse und Probleme. 33 Abb. VII, 174 S. 1968 Geb. DM 64,—; US $26.30   ISBN 3-540-04295-4

12   K. Scheffler; H. B. Stegmann: Elektronenspinresonanz. Grundlagen und Anwendung in der organischen Chemie. 145 Abb. VIII, 506 S. 1970. Geb. DM 148,—; US $60.70 ISBN 3-540-04984-3

13   Ch. Grundmann; P. Grünanger: The Nitrile Oxides. Versatile Tools of Theoretical and Preparative Chemistry. 1 fig. VIII, 242 pp. 1971 Cloth DM 98,—; US $40.20 ISBN 3-540-05226-7

14   M. Schlosser: Struktur und Reaktivität polarer Organometalle. Eine Einführung in die Chemie organischer Alkali- und Erdalkalimetall-Verbindungen. 29 Abb. XI, 187 S. 1973 Geb. DM 78,—; US $32.00 ISBN 3-540-05719-6

Preisänderungen vorbehalten

Springer-Verlag   Berlin   Heidelberg   New York